Hot Topics in Infection and Immunity in Children VI

Volume 659

A Continuation Order Plan is available for this series. A continuation order will bring delivery of each new volume immediately upon publication. Volumes are billed only upon actual shipment. For further information please contact the publisher.

Adam Finn · Nigel Curtis · Andrew J. Pollard
Editors

Hot Topics in Infection
and Immunity in Children VI

 Springer

Editors

Adam Finn
Institute of Child Life and Health
UBHT Education Centre
University of Bristol
Upper Maudlin Street
Bristol
UK BS2 8AE
adam.finn@bristol.ac.uk
a.finn@sheffield.ac.uk

Nigel Curtis
Royal Children's Hospital
Department of Paediatrics
University of Melbourne
Parkville VIC 3052
Australia
nigel.curtis@rch.org.au

Andrew J. Pollard
University of Oxford
Level 4, John Radcliffe
    Hospital
Oxford
UK OX3 9DU
andrew.pollard@paediatrics.ox.ac.uk

ISSN 0065-2598
ISBN 978-1-4419-0980-0   e-ISBN 978-1-4419-0981-7
DOI 10.1007/978-1-4419-0981-7
Springer New York Dordrecht Heidelberg London

Library of Congress Control Number: 2009938189

Printed on acid-free paper

Springer is part of Springer Science+Business Media (www.springer.com)

Preface

Each of the chapters in this book is based on a lecture given at the sixth "Infection and Immunity in Children" course, held at the end of June 2008 at Keble College, Oxford.

Thus, it is the sixth book in a series that provides succinct and readable updates on just about every aspect of the discipline of Paediatric Infectious Diseases.

The seventh course (29th June–1st July 2009) has a new and topical programme delivered by top-class speakers, and a seventh edition of this book will duly follow.

As we send this edition off for publication, the news of fatal cases of H1N1 swine flu in Mexico with spread into other countries around the globe is coming through, amidst fears of a global pandemic. By the time the book is being read, the size and severity of the problem will have become much clearer. For now, the outbreak serves as yet another reminder of the enduring importance of infectious diseases in human health and global prosperity and the need to continue studying them with all the epidemiological, clinical, and laboratory tools at our disposal.

We hope this book will provide a further useful contribution to the materials available to trainees and practitioners in this important and rapidly developing field.

UK, Australia   Adam Finn,
                                                                      Nigel Curtis,
                                                                  Andrew J. Pollard

Acknowledgments

We thank all the contributors who have written chapters for this book, which are all based on lectures delivered at the 2008 Infection and Immunity in Children course. We are very grateful to the staff of Keble College, Oxford, UK where the course was held.

Sue Sheaf administers and runs the course. As is the rule for people who organize extremely well, she creates the illusion that it is easy and straightforward, but we know it really is not, and she deserves enormous thanks from the organizers as well as all the speakers and delegates who have attended the Oxford IIC over recent years. Once again, Lorraine Cantle has administered the production of this book, obtaining the chapters from the authors – some of whom need more persuasion and reminding than others – and working with the editors to generate the finalized product for publication. We thank her for this patient work and gratefully share with her the credit for the finished object.

We thank the European Society for Paediatric Infectious Diseases for continuing practical and financial support for this series of courses. In particular, we draw attention to the ESPID bursaries towards the costs of many young delegates' attendance. We also acknowledge the recognition given to the course by the Royal College of Paediatrics and Child Health.

Finally, we are grateful to the several pharmaceutical industry sponsors who continue generously to offer unrestricted educational grants towards the budget for the meeting.

Contents

Contributors

Aditya H. Gaur Department of Infectious Diseases, St. Jude Children's Research Hospital, Memphis, Tennessee, USA

Alessandro R. Zanetti Department of Public Health-Microbiology-Virology, Faculty of Medicine, University of Milan, Milan, Italy

Alison J. Cody Department of Zoology, University of Oxford, South Parks Road, Oxford OX1 2PS, UK

Andrew Cant Paediatric Immunology & Infectious Diseases, Newcastle General Hospital, Westgate Road, Newcastle upon Tyne NE4 6BE, UK

Anne A. Gershon Department of Pediatrics, Columbia University College of Physicians and Surgeons, New York, NY 10032, USA

Aubrey Cunnington Immunology Unit, London School of Hygiene and Tropical Medicine, London, UK

B. Keith English The University of Tennessee Health Science Center, Memphis, Tennessee, USA

Cassandra Moran Department of Pediatrics, Duke University Medical Center, Durham, NC 27710, USA

Daniel Shouval Department of Medicine, Liver Unit, Hadassah-Hebrew University Hospital Jerusalem, Israel

Danny Benjamin Jr, MD Ph.D MPH Division of Quantitative Sciences, Duke Clinical Research Institute, Duke University Pediatrics, Durham, NC 27715, USA

David A. Kaufman Division of Neonatology, Department of Pediatrics, University of Virginia Health System, Charlottesville, VA 22908, USA

Deirdre Kelly The Liver Unit, Birmingham Children's Hospital, Birmingham B4 6NH, UK

Elizabeth Molyneux Paediatric Department, College of Medicine, P/Bag 360, Blantyre, Malawi

Frances M. Colles Department of Zoology, University of Oxford, South Parks Road, Oxford OX1 2PS, UK

H.K. Brand Department of Pediatrics, Radboud University Nijmegen Medical Center, Nijmegen, The Netherlands

Hermione Lyall Imperial College Healthcare NHS Trust, St. Mary's Hospital, London W2 1NY, UK

Koen Van Herck Centre for the Evaluation of Vaccination, Vaccine & Infectious Disease Institute, Faculty of Medicine, University of Antwerp, Antwerp, Belgium; 4Research Foundation – Flanders (FWO), Brussels, Belgium

Kondwani Kawaza Paediatric Department, College of Medicine, P/Bag 360, Blantyre, Malawi

Marc Tebruegge Department of Paediatrics, The University of Melbourne; Infectious Diseases Unit, Department of General Medicine; Microbiology & Infectious Diseases Research Group, Murdoch Children's Research Institute: Royal Children's Hospital Melbourne, Parkville, VIC 3052, Australia

Martin C.J. Maiden Department of Zoology, University of Oxford, South Parks Road, Oxford OX1 2PS, UK

Nigel Curtis Department of Paediatrics, The University of Melbourne; Infectious Diseases Unit, Department of General Medicine; Microbiology & Infectious Diseases Research Group, Murdoch Children's Research Institute: Royal Children's Hospital Melbourne, Parkville, VIC 3052, Australia

P.W.M. Hermans Department of Pediatrics, Radboud University Nijmegen Medical Center, Nijmegen, The Netherlands

Pierre Van Damme Faculty of Medicine, Centre for the Evaluation of Vaccination, Vaccine & Infectious Disease Institute, University of Antwerp, Antwerp, Belgium

R. de Groot Department of Pediatrics, Radboud University Nijmegen Medical Center, Nijmegen, The Netherlands

Samuel K. Sheppard Department of Zoology, University of Oxford, South Parks Road, Oxford OX1 2PS, UK

Sanjay Patel Department of Paediatric Infectious Diseases, St Mary's Hospital, London W2 1NY, UK

Simon J. Glover College of Medicine, University of Malawi, Blantyre, Malawi; Malawi-Liverpool-Wellcome Trust Clinical Research Programme, Blantyre, Malawi

Stephen J. Rogerson Department of Medicine (RMH/WH), Post Office Royal Melbourne Hospital, Parkville, VIC 3050, Australia

Theresa Cole Paediatric Immunology & Infectious Diseases, Newcastle General Hospital, Westgate Road, Newcastle upon Tyne NE4 6BE, UK

Yamikani Chimalizeni Paediatric Department, College of Medicine, P/Bag 360, Blantyre, Malawi

Infections in the Immunocompromised

Andrew Cant and Theresa Cole

1 Introduction

Infections in the immunocompromised differ significantly from those in the immunocompetent. They can be more serious, more often life threatening, more difficult to diagnose and are caused by more unusual organisms. Children can be immunocompromised for a variety of reasons and the numbers, worldwide, are growing.

2 Types of Immunodeficiency and Infection

Globally, the most common causes for immunocompromise are acquired. Severe malnutrition and the spread of HIV continue to be major causes of immunocompromise in the developing world. The development of more intensive chemotherapy and immunosuppressive drugs for malignancy and inflammatory diseases continues and the number of patients undergoing solid organ transplant (SOT) or haemopoietic stem cell transplant (HSCT) is also rising. Primary immunodeficiencies (PIDs) are also being increasingly recognized. Defects in humoral immunity make up the largest group of PIDs, at around 50%, with 20–30% being combined humoral and cellular deficiencies.

Different types of immunodeficiency or immunosuppression affect different aspects of the immune system and so predispose to different types of infection: bacterial, viral, fungal or protozoal. Understanding which part of the immune system is compromised helps identify the most likely infecting agent.

T lymphocyte defects predispose to infections such as life-threatening cytomegalovirus (CMV) pneumonitis or disseminated Epstein Barr virus (EBV) infection. These defects also present a significant risk of serious infection with

A. Cant (✉)
Paediatric Immunology & Infectious Diseases, Newcastle General Hospital, Westgate Road, Newcastle upon Tyne NE4 6BE, UK
e-mail: andrew.cant@nuth.nhs.uk

A. Finn et al. (eds.), *Hot Topics in Infection and Immunity in Children VI*, Advances in Experimental Medicine and Biology 659, DOI 10.1007/978-1-4419-0981-7_1, © Springer Science+Business Media, LLC 2010

cryptosporidium and *Aspergillus* species. B lymphocyte defects can result in infections by bacteria such as *Streptococcus pneumoniae, Haemophilus influenzae* and *Staphylococcus aureus,* as well as echo virus and protozoa such as *Giardia,* whereas phagocytic defects will result in infections by *Staphylococcus, Pseudomonas* and *Aspergillus* species.

When considering infections in those undergoing HSCT, it is also important to note that there is a recognized sequence of risk for different infections at different times after HSCT, which equates with different aspects of the immune system compromise at these times (Fig. 1). Within the first 30 days, when the patient is neutropenic, there is a significant risk of infection by both Gram-negative and Gram-positive bacteria, along with herpes-simplex virus. Between 30 and 90 days after transplant, when T-cell immunity is still limited, there is a rise in the numbers of fungal and CMV infections. Later infections are more commonly caused by varicella zoster virus (VZV) or *S. pneumoniae.*

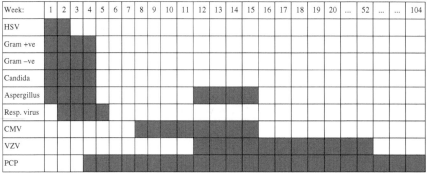

Week:	1	2	3	4	5	6	7	8	9	10	11	12	13	14	15	16	17	18	19	20	...	52	...	...	104
HSV																									
Gram +ve																									
Gram –ve																									
Candida																									
Aspergillus																									
Resp. virus																									
CMV																									
VZV																									
PCP																									

HSV: herpes simplex virus, Gram +ve: Gram positive bacteria, Gram –ve: Gram negative bacteria, CMV: cytomegalovirus, VZV: varicella zoster virus, PCP: *Ppneumocysitis jiroveci* pneumonia

Fig. 1 Diagrammatic representation of timing of infections after HSCT

Whatever the cause of the immunocompromise, possible infection requires a different approach to investigation and management from those in an immunocompetent child.

3 Site of Infection

Although an understanding of the type of immunocompromise is helpful to predict the likely organism, infection in the immunocompromised also needs to be considered by the system affected.

Infections can occur in any system of the body, but the respiratory system and gastrointestinal tract are especially vulnerable, as they have large surface areas and their barrier defences are, of necessity, compromised by the need to transport oxygen and nutrients, respectively. Disseminated viral and fungal infections are another important risk, whilst central venous catheter (CVC) infections also constitute a frequent problem in the immunocompromised. Each of these will be discussed in turn.

4 Respiratory Tract

The respiratory tract can be exposed to a wide variety of different organisms. *Pneumocystis jiroveci* pneumonia (PCP), CMV and *Aspergillus* are particularly important and well recognized sources of infection in the immunocompromised host; however, other significant pathogens have more recently been identified. These include respiratory syncytial virus (RSV); influenza; parainfluenza; adenovirus; picornaviruses; measles; human metapneumovirus; cocavirus; Coronaviruses NL63, and HKU1 and polyomaviruses WU and K1.

Pneumonitis and bronchiolitis are the most common presentations of respiratory infection, but lobar pneumonia may also occur. A defective immune/inflammatory response means that patients may have few respiratory symptoms, so there should be a low threshold for investigation. In one study where broncho-alveolar lavage (BAL) was performed in 69 children with immunodeficiency pre-HSCT, pathogens were isolated in 26 of these, six of whom were asymptomatic. PCP and bacteria were the most commonly identified organisms, followed by parainfluenza virus, CMV, RSV, influenza B and human herpes virus-6 (HHV6) (Slatter et al., 2007).

Accurate diagnosis depends on collecting the right samples and using appropriate diagnostic techniques. These include throat swabs, nasopharyngeal aspirates (NPA), BAL fluid and even lung biopsy, as deemed appropriate. Samples must be sent to look specifically for bacteria, fungi and viruses. Some respiratory pathogens will not be isolated from the upper respiratory tract; for example, PCP will not be identified on NPA, whilst other organisms found on NPA may not be found in the lower respiratory tract. This highlights the importance of BAL as a diagnostic procedure. Lung biopsy may be particularly important in the diagnosis of fungal infection, especially when there is a negative BAL in patients with persistent signs, symptoms or chest x-ray changes.

Diagnosis may require culture of organisms (bacteria, mycobacteria or viruses), immunofluorescence (viruses), polymerase chain reaction (bacteria, viruses and fungi) or antigen testing (e.g. galactomannan for *Aspergillus*). Serological testing is often ineffective, as immunodeficient children may not mount an antibody response or may be receiving intravenous immunoglobulin (IVIG), which will make results impossible to interpret. It is important to know what tests are available in your local laboratory. Discussion with the local microbiologist or virologist is essential to ensure the right samples are sent for appropriate investigations, so as not to miss a serious infection.

High resolution computerized tomography (HRCT) of the chest is more sensitive than chest x-ray, aiming to classify a disease as interstitial, airway or involving airspace, which may aid diagnosis.

4.1 *Pneumocystis jiroveci* Pneumonia (PCP)

PCP has historically been associated with HIV but is also a significant cause of morbidity in other groups of immunocompromised patients, particularly those with haematological malignancies, brain tumours requiring prolonged courses of steroids, prolonged neutropaenia or lymphopaenia, and those undergoing HSCT. Therefore PCP prophylaxis is important, as recommended by a recent Cochrane review (Green et al., 2007). This treatment is generally in the form of cotrimoxazole given three times per week. In children that cannot tolerate cotrimoxazole, either dapsone or aerosolized pentamidine can be used.

P. jiroveci infection commonly presents with tachypnoea, non-productive cough and fever, but the severity can vary. There is usually a sub-acute diffuse pneumonitis and chest x-ray changes can be subtle. These often take the form of bilateral diffuse interstitial changes, although lobar, miliary or nodular changes can be seen. HRCT may show ground glass attenuation, consolidation, nodules, thickening of interlobular septa and thin walled cysts. Mortality ranges between 5 and 40%, if treated, but can reach nearly 100% if left untreated.

Identification of PCP can be difficult. Definitive diagnosis depends on identifying the organism in respiratory tract secretions or lung tissue, usually from tracheal secretions, bronchial secretions or from lung biopsy. More recently, PCR technology has been developed for identifying PCP from secretions. In a review of children diagnosed with severe combined immune deficiency (SCID) treated at a supra-regional center, 10 out of 50 were identified as having PCP. One was diagnosed on BAL prior to transfer to the supra-regional center, one was diagnosed on nasopharyngeal secretions and BAL, seven were diagnosed on BAL alone, and in one diagnosis was not made until lung biopsy was performed (Berrington et al., 2000).

Recommended first line PCP treatment is high dose cotrimoxazole. This can, however cause a number of adverse effects, for example, neutropenia, anaemia, renal dysfunction, rash, vomiting and diarrhea. Those that cannot tolerate cotrimoxazole or those that have not improved after 5–7 days of treatment should be changed to a different agent. Choices include pentamidine, atovaquone, clindamycin/primaquine or dapsone, but experience with these agents in children is limited.

Corticosteroids should be given as an adjunctive therapy in moderate and severe PCP. A number of studies have shown a reduction in acute respiratory failure, decreased need for ventilation and decreased mortality (Sleasman et al., 1993; Bye et al., 1994; McLaughlin et al., 1995). A recent Cochrane review supports the use of corticosteroids in HIV-infected patients with PCP, especially in those with substantial hypoxaemia (Briel et al., 2006).

4.2 Viruses

A wide variety of respiratory viruses will also cause significant morbidity and mortality in the immunocompromised. Measles is an important respiratory pathogen in the immunocompromised host and it must be remembered that the typical rash may not develop. Mortality can be high, especially amongst patients with leukaemia and those undergoing HSCT. A prospective multi-center review of patients undergoing HSCT found direct RSV-associated mortality to be 17.4%, and mortality directly attributable to influenza A to be 15.3% (Ljungman et al., 2001). Respiratory viruses often present with non-specific symptoms but progress to a significant lower respiratory tract infection. Chest x-ray will often show a pneumonitis picture with diffuse interstitial changes. HRCT may show peri-bronchial thickening and ground glass attenuation without consolidation in a lobular distribution. Diagnosis requires identification of the organism from respiratory secretions. This may be possible on nasopharyngeal secretions or throat swab but may require more invasive testing, such as bronchoscopy and BAL. Laboratory techniques include immunofluorescence, PCR and viral culture. Treatment is mainly supportive, but specific treatment options are evolving, making rapid and accurate diagnosis increasingly important. Appropriate isolation and infection-control measures are essential to prevent transmission between immunocompromised patients, as these viruses can be easily spread. One UK study in a HSCT unit identified 10 cases of RSV over one winter season, and eight of the nine RSV strains that could be tested by molecular methods were found to be identical (Taylor et al., 2001).

Specific treatments for RSV infection include ribavirin and RSV monoclonal antibody (palivizumab). Ribavirin can be given orally, intravenously or via inhalation; however, the aerosolized route has been used most frequently for RSV infection. Historically, pooled hyperimmune RSV immunoglobulin has been proposed as an additional treatment, but this has been superseded by the anti-RSV monoclonal antibody, palivizumab. Combinations of inhaled ribavirin and intravenous palivizumab have shown encouraging results. Palivizumab has been shown to be safe and well tolerated in patients undergoing HSCT, with a suggestion of better outcome (improved survival) when compared to ribavirin alone (Boeckh et al., 2001; Chavez-Bueno et al., 2007).

There are two groups of drugs available for the treatment of influenza – namely the adamantanes (effective against influenza A, e.g. rimantadine) and the neuraminidase inhibitors (effective against both influenza A & B, e.g. oseltamivir). In recent years, there has developed increasing resistance to adamantanes. The neuraminidase inhibitors have been shown to reduce the duration of illness by one day when given to an immunocompetent host within 48 h of onset of symptoms. Although there are few data on the benefit of treating influenza in an immunocompromised patient with a neuraminidase inhibitor, their use appears sensible and safe.

There is, thus far, no specific treatment available for rhinovirus, coronavirus or human metapneumovirus. Ribavirin has been proposed as a treatment for parainfluenza virus infection but evidence, so far, of benefit is disappointing. Although there is little clinical data on the use of ribavirin for measles pneumonitis, it does have in vitro activity and therefore, due to the high level of mortality with this condition, should be considered.

A review of respiratory viral infection in children with primary immune deficiencies in a HSCT unit found 22 of 73 patients admitted for HSCT had respiratory viral infection. Of these, 11 had paramyxoviruses (RSV or parainfluenza I–IV), and were treated with aerosolized ribavirin and IVIG. Five of these patients also received nebulized immunoglobulin and corticosteroid. Three of these five survived, compared to two out of the six who did not receive nebulized treatment. It was concluded that the nebulized treatment was well tolerated and could be a useful adjunctive therapy (Crooks et al., 2000).

In children who have undergone HSCT, infection and inflammation can become inextricably interwoven to generate pneumonitis. In this case, in addition to the need for anti-infective agents, immunomodulation will be required through agents such as steroids, IVIG and anti-tumour necrosis factor monoclonal antibodies.

5 Gut

The gastrointestinal tract is also exposed to a wide variety of organisms and viruses which are of particular concern in the immunocompromised child, notably enteroviruses, adenovirus, rotavirus, caliciviruses, but also protozoa, mainly *Cryptosporidium* and *Giardia*. Presentation is most commonly with diarrhea and vomiting, which may protracted. *Cryptosporidium* can also be responsible for ascending cholangitis and liver disease. In some cases, identification of the causative organism can be difficult. Culture may be required to identify some viruses. PCR can be useful, for example, for adenovirus and is more sensitive than microscopy alone in detecting *Cryptosporidium*.

Prevention of transmission between immunocompromised patients is essential. There must be strict adherence to infection-control policies to prevent hospital wards from becoming sources of infection. One study looking at the extent of gastroenteric virus contamination in a pediatric-primary immunodeficiency ward and a general pediatric ward found viruses on 17 and 19% of environmental swabs, respectively. Interestingly, these were contaminating objects used by parents rather than staff – for example the parents' room television, the parents' toilet tap and the microwave used by parents on the pediatric-primary immunodeficiency ward (Gallimore et al., 2008). This highlights the importance of ensuring that parents and visitors, as well as staff, comply with hand washing and infection control measures.

5.1 Viruses

Rotavirus infection, which is usually relatively mild and self-limiting in the immunocompetent, can lead to persistent vomiting and diarrhea and, if untreated, severe malnutrition, in the immunocompromised. It can be identified in stool by using enzyme immunoassay and may also be identified on electron microscopy. There is no specific treatment. Fluid and electrolyte management is important. Orally administered immunoglobulin has been used in some cases.

Caliciviruses, namely Noroviruses and Sapoviruses can also cause significant problems in the immunocompromised. Symptomatic infection and virus shedding can be prolonged; for example, one case report of a child undergoing HSCT for cartilage hair hypoplasia demonstrated Norovirus shedding for 156 days following transplant, during the period of immune reconstitution. The child was symptomatic throughout this time (Gallimore et al., 2004). Again, there is no specific treatment but meticulous management of fluids, electrolytes and nutritional support is essential, allowing time for immune reconstitution and consequent viral clearance.

Adenovirus will be discussed in more detail in the section on disseminated infection in the immunocompromised.

5.2 Cryptosporidium

Cryptosporidium species are oocyst-forming protozoa that cause watery diarrhea which can result in severe dehydration and even death, if not treated. Disease is normally confined to the gastrointestinal tract, but there is a risk of biliary tree, pulmonary or even disseminated disease in the immunocompromised. Infection may be diagnosed on identification of oocysts by microscopy. Enzyme immunoassays have also been used and PCR, too, can be helpful.

Treatment of *Cryptosporidium* infection can be difficult and a number of agents have been proposed, including nitazoxanide, paromomycin, rifabutin and the macrolides. Evidence is limited but a recent review has indicated that nitazoxanide may reduce parasite load and therefore be useful (Abubakar et al., 2007). In the authors' experience, azithromycin and nitazoxanide are safer options in post-HSCT patients, as paromomycin has been associated with significant hearing loss, particularly when given with ciclosporin. Supportive care remains essential. In those with HIV, anti-retroviral therapy, with its associated improvement in CD4 count, can result in improvement in the *Cryptosporidium* infection.

5.3 Giardia

Giardia intestinalis is a flagellate protozoan that exists in trophozoite or cyst forms. The cysts are the infective form. Children with humoral immunodeficiencies are particularly at risk of chronic symptomatic infection, with foul-smelling stool, abdominal distension and anorexia. Cysts may be identified on stool microscopy or by using immunofluorescent antibody testing. Treatment is with metronidazole, tinidazole or nitazoxanide. It may be necessary to use combination therapy in the immunocompromised if they have failed to respond to single-agent treatment.

6 Disseminated Infection

Disseminated viral infection in the immunocompromised is of particular concern. The most significant culprits are adenovirus and members of the human herpes virus: CMV, EBV, HHV6, HSV and VZV. These can affect the lungs, gastrointestinal tract and brain, resulting in a variety of symptoms. Reactivation of latent herpes viral infection is more common than primary infection after SOT or HSCT. Investigation using PCR techniques allows early diagnosis and quantification of viral load, and is now possible for adenovirus, CMV, EBV and HHV6. Prophylaxis to prevent CMV and HSV reactivation is used for children undergoing HSCT and many SOTs. Surveillance in high-risk patients enables pre-emptive treatment to be given before damaging disease occurs. Treatment will depend on the causative virus.

6.1 Adenovirus

Adenovirus is usually responsible for relatively minor upper respiratory tract or gastrointestinal infection but can result in life-threatening pneumonia, meningitis, encephalitis and disseminated disease in the immunocompromised. Those most at risk are patients who receive allogeneic bone marrow transplant, those with active graft versus host disease and those who receive total body irradiation. There are a number of different species of adenovirus, and these are divided into serotypes, some of which are primarily associated with the respiratory tract, while others have a predilection for the gastrointestinal tract. Young children are particularly vulnerable, as they often carry adenovirus in their gastrointestinal tract, predisposing them to reactivation and dissemination when they become immunocompromised. In view of this, screening can be important in the immunocompromised and adenovirus is usually identified in urine, stool, or sometimes respiratory secretions prior to being identified in blood. A study of 132 patients undergoing HSCT were screened for adenovirus in stool, urine, on throat swab and in peripheral blood during the post transplant period. 27% had a positive adenoviral PCR on at least one screening test, but this was not associated with clinical signs unless it was detected in peripheral blood and, even then, there was a median delay of 3 weeks from first detection of adenovirus until the patient demonstrated clinical signs. In one study, mortality was as high as 82% in those with adenovirus detected on peripheral blood. This highlights the importance of early recognition and consideration of pre-emptive use of antivirals (Lion et al., 2003).

Successful treatment of adenovirus infection has so far been limited. The most widely used agents are cidofovir or ribavirin, which may be given together with IVIG. Although cidofovir has potent nephrotoxic effects, these can be greatly reduced by the concurrent use of intravenous hyperhydration and probenecid. Cidofovir has been shown to be more effective in adenovirus and is now considered the best first-line treatment. Data on the clinical effectiveness of ribavirin in adenoviral infections are more conflicting. In vitro data suggest that ribavirin alone has activity against subgenus C serotypes. In a post-HSCT patient with adenoviral infection, immune suppression should be reduced as much as possible, as T-cell immune reconstitution is very important for viral elimination.

6.2 Cytomegalovirus

CMV infection is often asymptomatic in the immunocompetent; however, in the immunocompromised it can lead to pneumonia, colitis and retinitis. CMV persists in a latent form after primary infection and can result in reactivation in someone who later becomes immunosuppressed – for example, when undergoing HSCT. CMV can be identified from respiratory secretions, urine and blood. As with adenovirus, PCR screening may be useful in identifying the virus before a child becomes symptomatic, especially in cases where reactivation is likely with immunosuppression. Treatment is usually with intravenous ganciclovir, with foscarnet or cidofovir as second-line treatment. Oral valganciclovir is very well absorbed and is also now an option for treatment. Foscarnet has also been used in cases of children undergoing HSCT to avoid the myelosuppressive effects of ganciclovir. IVIG should be used alongside antiviral therapy. There has been one case report of ganciclovir- and foscarnet-resistant CMV being successfully treated with artesunate (Shapira et al., 2008). There is also interest in the new antiviral agent maribavir for resistant CMV.

6.3 Epstein Barr Virus

EBV is associated with lymphoproliferative disorders in the immunocompromised. Replication of EBV in B cells is usually inhibited by natural killer cells, antibody-dependent cell cytotoxicity and T-cell cytotoxic responses. Therefore, children with cellular immune deficiencies are at risk of uncontrolled lymphoproliferation. Those at particular risk are children who are transplant recipients, both SOT or HSCT, and those with HIV. EBV can be detected in blood by PCR and viral load can be monitored. Alongside monitoring of the virus, it is important to monitor for signs of lymphoproliferation, both clinically and biochemically. Biopsy of suspicious lesions is often needed to make a diagnosis.

EBV infection requires treatment if it causes B lymphoproliferation or post-transplant lymphoproliferative disease (PTLD). This may take the form of the anti-CD20 monoclonal antibody rituximab, chemotherapy or radiotherapy. Decreasing immunosuppression whenever possible in a post-transplant patient is very important. More recently there have been encouraging results from work with cytotoxic T-cell therapy in PTLD. This involves the infusion of EBV-specific cytotoxic T lymphocytes (CTLs) generated from EBV sero-positive blood donors. In one recent multi-center study, 33 patients who had failed conventional therapy were recruited and monitored for response: 14 patients achieved complete remission while three showed a partial response (Haque et al., 2007).

6.4 Human Herpes Virus 6

Primary HHV6 infection in the immunocompetent host leads to the typical clinical picture of roseola or a non-specific febrile illness. The virus remains latent after primary infection and therefore, similar to CMV, can reactivate in immunocompromised states. The importance of HHV6 as a pathogen in the immunocompromised is probably underestimated, and many labs do not screen for infection; thus, many infections may not be recognized. HHV6 can cause fever, rash, hepatitis, pneumonia and encephalitis, as well as bone marrow suppression. HHV6 also appears to have synergistic effects and interactions with other infectious agents, such as CMV, adenovirus and fungi. It can be identified and quantified on blood samples by PCR. Treatment, where necessary, is with intravenous ganciclovir or foscarnet.

6.5 Varicella Zoster Virus

Primary varicella infection results in chickenpox, a common and generally self-limiting childhood illness. In the immunocompromised, there is a significant risk of both primary or reactivated disease becoming disseminated. This is particularly associated with T lymphocyte defects. VZV is the second most common cause of viral pneumonitis in children with AIDS. It should be remembered that fatal VZV infection has been reported in cases where the only immunosuppressant medication has been corticosteroids at a dose of 1 mg/kg/day of prednisolone for 2 weeks. The virus can be identified from vesicular fluid. Treatment is usually in the form of intravenous aciclovir, but, oral valaciclovir is a useful alternative in older children.

An important area to consider in relation to VZV infection is that of post-exposure prophylaxis. Although long-term prophylaxis for VZV is not usually recommended, post-exposure prophylaxis in non-immune immunocompromised children is important. Two options are available. The most widely used is varicella zoster immunoglobulin (VZIG). However, due to a shortage of VZIG a few years ago, oral aciclovir was reconsidered and has been shown to be effective. It must be remembered, however, that aciclovir has low bioavailability when given orally and requires multiple daily dosing. It may be more appropriate to consider the oral pro-drug valaciclovir, which has been shown to be effective and well tolerated (Nadal et al., 2002). Further work to clarify the best prophylactic and pre-emptive treatment regimens is needed.

7 Fungal Infections

Fungal infections must be considered in specific circumstances; for example, in those who are neutropenic (where risk increases exponentially with duration of neutropenia), those on steroids and those with graft versus host disease. *Candida* and *Aspergillus* are of particular interest in children who have undergone HSCT.

Symptoms that should raise the suspicion of fungal infection are persistent fevers unresponsive to antibiotics, skin nodules, chest pain and radiological evidence of infection crossing tissue planes. *Candida* is most commonly associated with CVC infection but can also cause disseminated disease. *Aspergillus* infection can have an insidious onset, frequently affecting the respiratory tract but then spreading to involve other areas such as the spine and intracranial cavity. Investigation and diagnosis remain difficult and may require antigen testing, PCR, cross-sectional imaging and biopsy of suspicious lesions/areas.

7.1 Candidiasis

Persistent mucocutaneous candidiasis is seen in patients with defects in T-cell function and may be a presenting feature for HIV infection or primary immune deficiency. Disseminated infection can involve almost any organ or any anatomical site and can be rapidly fatal. It is a particular concern in patients with CVC, especially those receiving multiple infusions and/or parenteral nutrition. There are a number of different *Candida* species that can result in disseminated infection. *Candia albicans* is the most common but *C. parapsilosis*, *C. glabrata*, *C. tropicalis* and *C. krusei* are increasingly common (Fig. 2). Diagnosis may be difficult, as blood cultures are not always positive. However, identification can be made by microscopy of biopsy specimens. Suspicious lesions, which are often found in organs such as the liver, kidney, spleen and brain, are best identified by cross-section imaging. PCR techniques have been developed, as well as detection of antigen from the fungal cell wall (Mannan). However, these techniques are not as yet wholly reliable. There are a number of agents available for treatment, including amphotericin B, caspofugin or an azole, such as voriconazole. Prolonged treatment is usually required and if there is a CVC

Species	Potential resistance
C. albicans	Resistance remains rare
c. glabrata	Fluconazole resistance in up to 20% cases, decreased susceptibility to other azoles; some cases of amphotericin B resistance
C. parapsilosis	Decreased susceptibility to caspofungin
C. tropicalis	
C. krusei	Resistant to ketoconazole and fluconazole; decreased susceptibility to itraconazole and amphotericin B
C. guillermondi	Decreased suscepatbility to fluconazole
C. lustaniae	Generally resistant to amphotericin B
C. rugosa	Resistant to fluconazole in up to 60% cases; resistant to voriconazole

Fig. 2 *Candida* species and potential resistance to antifungal agents

involved it should be removed. Recently, there have been increasing concerns about anti-fungal drug resistance.

7.2 Aspergillosis

Invasive *Aspergillus* infection in the immunocompromised usually involves lungs, sinuses, brain or skin and commonly crosses tissue planes. Less commonly, it can cause endocarditis, osteomyelitis, meningitis and infection around the eye or orbit. It can cause angio-invasion, resulting in thrombosis and, occasionally, erosion of the blood vessel wall, often with catastrophic hemorrhage as a consequence. There are a number of *Aspergillus* species that cause invasive disease. Most commonly it is due to *Aspergillus fumigatus,* but *A. flavus*, *A. terreus*, *A. nidulans* and *A. niger* are also responsible for invasive infection. Diagnosis can be challenging. Cross-sectional imaging is very important in identifying suspicious lesions. *Aspergillus* is infrequently identified from blood and is most commonly indicated from biopsy specimens. Galactomannan, a complex sugar molecule found in the cell wall of the *Aspergillus* species, may also be identified from blood and can be useful in aiding diagnosis. Treatment is usually with amphotericin B, voriconazole or caspofungin and requires a prolonged course. Surgical excision of fungal lesions may be required, especially if there are significant areas of necrotic tissue into which antifungal agents will not penetrate effectively.

There is also an important association between *Aspergillus* infection and building work on a hospital site. One study in an Italian hematology unit found three cases of proven Aspergillosis in patients with acute leukemia that coincided with renovation work on the hospital site and high levels of *A. fumigatus* in the corridors (Pini et al., 2008). This highlights the importance for high-risk patients (e.g. after HSCT) of sterile isolation in cubicles maintained at positive pressure with highly purified air. Extra attention must be paid to reducing exposure of immunocompromised patients when there is building work on any hospital site.

8 Central Venous Catheter Infections

Many immunocompromised children will have indwelling CVC for treatment, be this an external Broviac or Hickman line, or an internal Portacath. Although very beneficial they, unfortunately, provide a site for infection. Catheter-related blood stream infections can be serious and in some cases life-threatening. Clinical features of catheter-related blood stream infection can be very non-specific. Diagnosis is often made on identification of organisms from blood culture along with lack of focal infective symptoms/signs. Organisms causing CVC infection are often those that would be non-virulent normal flora in an immunocompetent host; for example, coagulase negative staphylococci, enterococci and viridans streptococci. However, mycobacterial CVC infections also occur (Hawkins et al., 2008), as do *Candida* CVC infections.

Prevention has to be the priority. Lines should be inserted under strict aseptic technique and, once in place, access should be by fully trained staff using aseptic technique. Local policies should be followed for accessing and flushing CVCs.

Historically, CVCs were often removed when infection was identified; however, many patients were left in the difficult situation of poor venous access and in need of a further general anaesthetic to replace the line. Many catheter-related blood stream infections can be treated with antibiotics, without requiring CVC removal. If there is clinical suspicion of catheter-related blood stream infection, antibiotics for both coagulase negative *Staphylococcus* and Gram negative organisms should be introduced. Once organisms are identified from blood culture, antibiotics can be tailored appropriately. Antibiotic "locks" can be used alongside systemic antibiotics to reduce colonization within the CVC. Antibiotic "locking" involves instilling 1–2 ml of concentrated antibiotic solution in to the CVC and leaving it for a pre-determined time before removal. Antibiotics used in studies to treat CVC colonization have included vancomycin, amikacin and minocycline. There is also limited evidence on the use of amphotericin locks. Studies have attempted to look at whether using locks alone or in combination with systemic antibiotics has benefits. The results are variable and, at this stage it must be concluded that locks are a useful adjunct to systemic treatment. There is not enough evidence to suggest they can be used alone in immunocompromised children with CVCs (Berrington and Gould 2001).

In an attempt to present CVC infection, antibiotic-impregnated CVCs have also been developed. A recent systematic review found significant reductions in catheter-related blood stream infections in heparin-coated or antibiotic-impregnated CVCs, when compared to standard CVCs, as well as those coated with chlorhexidine, silver sulphadiazine, or silver-impregnated. There were, however, some concerns about the development of antibiotic resistance and further study is required before recommendations can be made about the most appropriate CVC to be used (Gilbert and Harden, 2008).

It must be remembered that catheter-related blood stream infection can be life threatening and there should be a low threshold for removal of the CVC if there are signs of clinical deterioration on treatment or if blood cultures drawn from CVCs are repeatedly positive, despite ongoing appropriate antibiotic treatment. There is increased mortality associated with delayed catheter removal in *S. aureus* and fungal infections, and so removal must be considered urgent if these organs are isolated. The benefits of removing the CVC if Gram-negative organisms are identified is slightly more difficult to assess due to scarcity of data; however, it is likely that immediate removal does contribute to increased survival. In all infections the risk/benefit ratio of removing or retaining CVCs should be carefully considered.

9 Febrile Neutropenia

In children receiving treatment for malignancy, febrile neutropenia is a significant cause of morbidity and mortality. Over time, outcome has improved dramatically but it still remains a frequent reason for hospitalization. It has been shown that

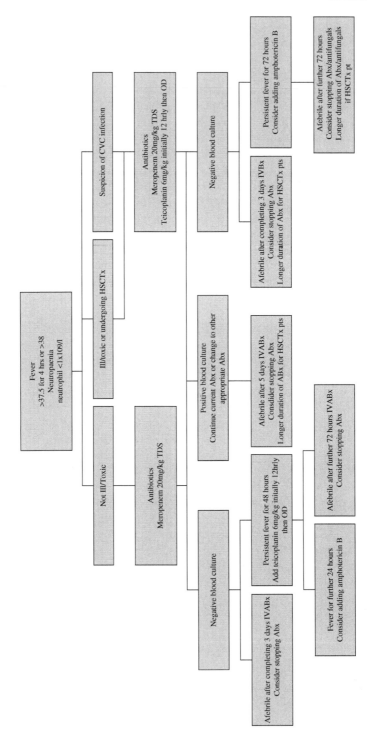

Fig. 3 Suggested protocol for febrile neutropaenia treatment

empiric use of antibiotics decreases mortality (Schmipff et al., 1971); hence, empiric antibiotics have become a standard part of treatment for children and adults with febrile neutropenia.

Fever with neutropenia in any immunocompromised child should be acted on promptly. However, exactly how this is defined and what is appropriate management varies widely. This was highlighted by a recent review of febrile neutropenia management in the United Kingdom Children's Cancer Study Group centers (Phillips et al., 2007). There was wide variation in the definition of fever (from persistent temperature higher than 37.5°C to a single reading of 39°C) and neutropenia (absolute neutrophil count $<1 \times 10^9$, $<0.75 \times 10^9$ or $<0.5 \times 10^9$). Empirical antibiotic regimes also varied greatly, including aminoglycosides plus a second agent (piperacillin based, cephalosporin or carbapenem), carbapenem alone or, in two cases, cefuroxime plus flucloxacillin and ciprofloxacin plus ceftazidime. Timing of the anti-fungal therapy was even more variable, in terms of when to start and the duration of empirical treatment. Some of this variation can be explained by variations in organisms isolated and antibiotic sensitivity from unit to unit, but this does not seem to account for all the differences in practice. Therefore, although local findings should influence presenting patterns, further work is required to devise a framework within which local policies that target specific patient populations and microbiological flora are implemented. A specimen protocol is shown in Fig. 3

10 Conclusions

Infections in immunocompromised children offer a variety of challenges in both diagnosis and management. Organisms that result in mild, self-limiting illness in an immunocompetent host can have catastrophic effects on an immunocompromised child. Signs and symptoms are often less specific and finding a causative organism can be more difficult. It is important to have a low threshold for thinking about infections and looking for them. Negative tests should not be taken to be reassuring if there is clinical suspicion and it may be necessary to look further and more closely. It is important to develop a good relationship with local microbiology and virology laboratories to aid this process. Once an infection is identified, it must be acted upon quickly as delay may be disastrous. Treatment of any infection in an immunocompromised child is likely to be more intense and prolonged than in a child with a fully functioning immune system. It is also important to consider prophylaxis for specific patient groups in specific situations (e.g. post HSCT) and each unit should have defined policies and guidelines to follow for these patients.

In summary, when dealing with an immunocompromised child, for whatever reason, when there is suspicion about infection, think early, look carefully and treat now!

References

Abubakar, I., Aliyu, S.H., Arumugam, C., Usman, N.K., & Hunter, P.R. (2007). Treatment of crypotspodiiosis in immunocompromised individuals: systematic review and meta-analysis. *Br J Clin Pharm*, (63), 387–393

Alexander, S.W., Mueller, B.U., & Pizzo, P. (2002). Infectious complications in children with cancer and children with Human Immunodeficiency Virus Infection. *Clinical Approach to Infection in the Compromised Host*. R. Rubin, L.S. Young (Eds.) pp. 441–464. New York: Kluwer Academic/Plenum Publishers

Andris, D.A., Krzywda, E.A., & Edmiston, C.E. (1998). Elimination of intraluminal colonisation by antibiotic lock in silicone vascular catheters. *Nutrition*, (14), 427–432

Baum, S.G. (2000). Adenovirus. *Mandell, Douglas and Bennet's Principles & Practice of Infectious Disease* (5th ed.) G.L. Mandell, J.E. Bennet, R. Dolin (Eds.) pp. 1624–1630. Philadelphia: Churchill Livingstone

Benfield, T., Atzori, C., Miller, R.F., & Helweg-Larsen J. (2008). Second line salvage treatment of AIDS-associated Pneumocystis jiroveci pneumonia: a case series and systematic review. *J Acquir Immune Defic Syndr*, 48(1), 63–67

Berrington, A. & Gould, K. (2001). Use of antibiotic locks to treat colonized central venous catheters. *J Antimicrob Chemother*, (48), 597–603

Berrington, J.E., Flood, T.J., Abinun, M., Galloway, A., & Cant, A.J. (2000). Unsuspected Pneumocystis carinii pneumonia at presentation of severe primary immunodeficiency. *Arch Dis Child*, (82), 144–147

Biron, K.K. (2006). Antiviral drugs for cytomegalovirus diseases. *Antiviral Res*, (71), 154–163

Boeckh, M., Berrey, M.M., Bowden, R.A., Crawford, S.W., Balsley, J., & Corey, L. (2001). Phase 1 evaluation of the respiratory syncytial virus specific monoclonal antibody Palivizumab in recipients of hematopoietic stem cell transplants. *JID*, (184), 350–354

Boeckh, M. (2006). Prevention of VZV infection in immunosuppressed patients using antiviral agents. *Herpes*, 13(3), 60–65

Briel, M., Bucher, H.C, Boscacci, R., & Furrer, H. (2006). Adjunctive corticosteroids for Pneumocystis jiroveci pneumonia in patients with HIV infection. *Cochrane Database Syst Rev*, 2006 Jul, (193), CD006150

Bye, M.R., Cairns-Bazarian, A.M., & Ewig, J.M. (1994). Markedly reduced mortality associated with corticosteroid therapy of Pneumocystis carinii pneumonia in children with acquired immune deficiency syndrome. *Arch Pediatic Adolesc Med*, (148), 638–641

Chakraborty, R. & Shingadia, D. (2006). Treating opportunistic infections in HIV infected children. Guidelines for the Children's HIV association (CHIVA)

Chavez-Bueno, S., Mejias, A., Merryman, R., Ahmad, N., Jafri, H.S., & Ramilo, O. (2007). Intravenous palivizumab and ribavirin combination for respiratory syncytial virus disease in high-risk paediatric patients. *Ped Inf Dis J*, 26(12), 1089–1093

Cherry, J.D. (1998). Adenoviruses. *Textbook of Pediatric Infectious Diseases* (4th ed.) R.D. Fegin, & J.D. Cherry (Eds.) pp. 1666–1684. Philadelphia: W.B. Saunders Company.

Crooks, B.N.A., Taylor, C.E., Turner, A.J.L., Osman, H.K.E., Abinun, M., Flood, T.J., & Cant, A.J. (2000). Respiratory viral infections in primary immune deficiencies: significance and relevance to clinical outcome in a single BMT unit. *Bone Marrow Transplant*, (26), 1097–1102

Drew, W.L., Miner, R.C., Marousek, G.I, & Sunwen, C. (2006). Maribavir sensitivity of cytomegalovirus isolates resistant to ganciclovir, cidofovir or foscarnet. *J Clin Virol*, (37), 124–127

Gallimore, C., Taylor, C., Gennery, A.R., Cant, A.J., Galloway, A., Xerry, J., Adigwe, J., & Gray, J.J. (2008). Contamination of the hospital environment with gastroenteric viruses: comparison of two pediatric wards over a winter season. *J Clin Microbiol*, 46(9), 3112–3115

Einsele, H., Reusser, P., Bornhauser, M., Kalhs, P., Ehninger, G., Hebart, H., Chalandon, Y., Kroger, N., Hertenstein, B., & Rohde, F. (2006). Oral valganciclovir leads to higher exposure to ganciclovir than intravenous ganciclovir in patients following allogeneoic stem cell transplantation. *Blood*, (107), 3002–3008

Gallimore, C., Lewis, D., Taylor, C., Cant, A., Gennery, A., & Gray, J. (2004). Chronic excretion of a norovirus in a child with cartilage hair hypoplasia(CHH). *J Clin Virol*, (30), 196–204

Garret Nichols, W., Peck Campbell, A.J., & Boeckh, M. (2008). Respiratory viruses other than influenza virus: impact and therapeutic advances. *Clin Micro Rev*, 21(2), 274–290

Gaynor, A.M., Nissen, M.D., Whiley, D.M., Mackay, I.M., Lambert, S.B., Guang, W.U., Brennan, D.C., Storch, G.A., Sloots, T.P., & Wang D. (2007). Identification of a novel polyomavirus from patients with acute respiratory tract infections. *PLoS Pathog*, 3(5), e64. doi:10.1371/journal.ppat.0030064

Green, H., Paul, M., Vidal, L., & Leibovici, L. (2007). Prophylaxis for Pneumocystis pneumonia (PCP) in non-HIV immunocompromised patients. *Cochrane Database Syst Rev*, 2007 Jul 18, (3), CD005590

Gilbert, R.E. & Harden, M. (2008). Effectiveness of impregnated central venous catheters for catheter related blood stream infection: a systematic review. *Curr Opin Infect Dis*, 21(3), 235–245

Haque, T., Wilie, G.M., Jones, M.M., Higgins, C.D., Urquhart, G., Wingate, P., Burns, D., McAulay, K., Turner, M., Bellamy, C., Amlot, P.L., Kelly, D., MacGilchrist, A., Gandhi, M.K., Swerdlow, A.J., & Crawford, D. (2007). Allogeneic cytotoxic T cell therapy for EBV positive posttransplantation lymphoproliferative disease: results of a phase 2 multicenter clinical trial. *Blood*, (110), 1123–1131

Hawkins, C., Qi, C., Warren, J., & Stosor, V. (2008). Catheter related bloodstream infections caused by rapidly growing nontuberculous mycobacteria: a case series including rare species. *Diagn Microbiol Infect Dis*, 61(2), 187–191

Hodge, D. & Puntis, J.W.L. (2002). Diagnosis, prevention and management of catheter related bloodstream infection during long term parenteral nutrition. *Arch Dis Child Fetal Neonatal Ed*, (87), F21–F24

Investigation of Blood Cultures (for organisms other than Mycobacterium species). Issue no: 5. issue date 09.08.05. Issued by: Standards Unit, Evaluations and Standard Laboratory. Reference no: BSOP 37i5

Lee, I & Barton, T.D. (2007). Viral respiratory tract infections in transplant patients. *Drugs*, 67(10), 1411–1427

Lenaerts, L. & Naesens, L. (2006). Antiviral therapy for adenovirus infections. *Antiviral Res*, (71), 172–180

Lion, T., Baumgartinger, R., Watzinger, F., Matthes-Martin, S., Suda, M., Preuner, S., Futterknecht, B., Lawitschka, A., Peters, C., Pötschegr, U., & Gadner, H. (2003). Molecular monitoring of adenovirus in peripheral blood after allogeneic bone marrow transplantation permits early diagnosis of disseminated disease. *Blood*, 102(3), 1114–1120

Ljungman, P., Ward, K.N., Crooks, B.N.A., Parker, A., Martine, R., Shaw, P.J., Brinch, L., Brune, M., De La Camara, R., Dehher, A., Pauksen, K., Russell, N., Schwarere, A.P., & Cordonnier, C. (2001). Respiratory virus infections after stem cell transplantation: a prospective study from the Infectious Diseases Working Party of the European Group for Blood and Marrow Transplantation. *Bone Marrow Transplant*, (28), 479–484

McLaughlin, G.E., Virdee, S.S., Schleien, C.L., Holzman, B.H., & Scott G.B. (1995). Effect of corticosteroids on survival of children with acquired immune deficiency syndrome and Pneumocystis carinii-related respiratory failure. *J Pediatr*, (126), 821–824

Nadal, D., Leverger, G., Sokal, E.M., Floret, D., Perel, Y., Leibundgut, K., & Weller S. (2002). In investigation of the steady-state pharmacokinetics of oral valaciclovir in immunocompromised children. *JID*, 186(Suppl 1), S123–130

Pfaller, M.A. & Diekema, D.J. (2007). Epidemiology of invasive candidiasis: a persistent public health problem. *Clin Micro Rev*, 20(1), 133–163

Phillips, B., Selwood, K., Lane, S.M., Skinner, R., Gibson, F., & Chisholm, J. (2007). Variation in policies for the management of febrile neutropaenia in United Kingdom Children's cancer Study Group centres. *Arch Dis Child*, (92), 495–498

Pickering, L.K., Baker, C.J., Long, S.S., & McMillan, J.A, (Eds.) (2006). *Red Book. Report of the Committee on Infectious Diseases.* (27th ed.). Elk Grove Village, IL: American Academy of Paediatrics.

Pini, G., Faggi, E., Donato, R., Sacco, C., & Fanci, R. (2008). Invasive pulmonary aspergillosis in neutropenic patients and the influence of hospital renovation. *Mycoses*, 51(2), 117–122

Schildgen, O., Müller, A., Allander, T., Mackay, I.M., Völz, S., Kupfer, B., & Simon, A. (2008). Human bocavirus: passenger of pathogen in acute respiratory tract infections? *Clin Micro Rev*, 21(2), 291–304

Schimpff, S., Slatterlee, W., & Young, V.M. (1971). Empiric therapy with carbenicillin and gentamicin for febrile patients with cancer and granulocytopenia. *NEJM*, (284), 1061–1065

Shankar, S.M. & Nania, J.J. (2007). Management of Pneumocystis jiroveci pneumonia in children receiving chemotherapy. *Paediatr Drugs*, 9(5), 301–309

Shapira, M.Y., Resnick, I.B., Chou, S., Neumann, A.U., Lurain, N.S., Stamminger, T., Caplan, O., Saleh, N., Efferth, T., Marschall, M., & Wolf, D. (2008). Artesunate as a potent antiviral agent in a patient with late drug-resistant cytomegalovirus infection after hematopoietic stem cell transplantation. *CID*, (46), 1455–1457

Slatter, M.A., Rogerson, E.J., Taylor, C.E., Galloway, A., Clark, J.E., Flood, T.J., Abinun, M., Cant, A.J., & Gennery, A.R. (2007). Value of bronchoalveolar lavage before haematopoietic stem cell transplantation for primary immunodeficiency or autoimmune diseases. *Bone Marrow Transplant Sep*, 40(6), 529–533.

Sleasman, J.W., Hemenway, C., Klein, A.S., & Barrett D.J. (1993). Corticosteroids improve survival of children with AIDS and Pneumocystis carinii pneumonia. *Am J Dis Child*, (147), 30–34

Soldatou, A. & Graham Davies, E. (2003). Respiratory virus infections in the immunocompromised host. *Paed Resp Rev*, (4), 193–204

Taylor, G.S., Vipond, I.B., & Caul, E.O., (2001). Molecular epidemiology of outbreak of respiratory syncytial virus within bone marrow transplant unit. *J Clin Microbiol*, 39(2), 801–803

Van Burik, J. & Weisdorf, D. (2000). Infections in recipients of blood and marrow transplantation. *Mandell, Douglas and Bennet's Principles & Practice of Infectious Disease*. (5th ed.) G.L. Mandell, J.E. Bennet, R. Dolin (Eds.) p. 3138. Philadelphia: Churchill Livingstone.

Walls, T., Shankar, A.G., & Shingadia, D. (2003). Adenovirus: an increasingly important pathogen in paediatric bone marrow transplant patients. *The Lancet*, Feb, (3), 79–86

Yee-Guardino, S., Gowans, K., Yen-Lieberman, B., Berk, P., Kohn, D., Wang, F.Z., Danziger-Isakov, L., Sabella, C., Worley, S., Pellet, P.E., & Goldfarb, J. (2008). Beta-herpesviruses in febrile children with cancer. *Emerg Infect Dis*, 14(4), 579–585

Host Biomarkers and Paediatric Infectious Diseases: From Molecular Profiles to Clinical Application

H.K. Brand, P.W.M. Hermans, and R. de Groot

1 Introduction

Infectious diseases are an important cause of death among children under the age of 5 (Stein et al., 2004). Most of these deaths are caused by preventable or curable infections. Limited access to medical care, antibiotics, and vaccinations remains a major problem in developing countries. But infectious diseases also continue to be an important public health issue in developed countries. With the help of modern technologies, some infections have been effectively controlled; however, new diseases such as SARS and West Nile virus infections are constantly emerging. In addition, other diseases such as malaria, tuberculosis, and bacterial pneumonia are increasingly resistant to antimicrobial treatment.

The physician who manages pediatric patients with infectious diseases is confronted with several related challenges. First, one should establish a specific diagnosis, preferably early in the course of disease. Despite improvements in culture and non-culture diagnostics, in many cases, the causative micro-organism remains unknown. Consequent delays in initiation of appropriate treatment can contribute to the emergence of antibiotic resistance.

A second challenge is to identify those patients most likely to develop severe disease. To date, physicians have little information on prognosis and likely disease outcome in the individual patient. It would be extremely useful to be able to identify patients at risk of more severe disease (e.g., secondary bacterial infection during viral respiratory tract infection), as such prediction could inform management decisions.

The third associated challenge is to select the most appropriate treatment strategy for an individual patient. While some patients require intensive support, others will recover without additional medication or supportive care. To date, few tools are available to monitor the course of disease after initiation of medical treatment.

R. de Groot (✉)

Department of Pediatrics, Radboud University Nijmegen Medical Center, Nijmegen, The Netherlands

e-mail: r.degroot@cukz.umcn.nl

A. Finn et al. (eds.), *Hot Topics in Infection and Immunity in Children VI*, Advances in Experimental Medicine and Biology 659, DOI 10.1007/978-1-4419-0981-7_2, © Springer Science+Business Media, LLC 2010

Biomarkers have been used for years to help clinical decision-making. C-reactive protein (CRP) is probably the best known marker used to monitor infection. Although useful, it does not reliably distinguish viral from bacterial infections. More recently-developed markers such as procalcitonin seem promising, with detectable rises early in the course of infection and high negative predictive value as seen in children with fever of unknown origin (van Rossum et al., 2004; Galetto-Lacour et al., 2003; Herd, 2007). However, this marker has also insufficient power to discriminate between viral and bacterial infections. Additionally, these conventional biomarkers for infectious diseases do not provide microorganism-specific prognostic information.

The completion of the Human Genome Project and the introduction of powerful DNA microarray chips and proteomic technologies in the mid-1990s have created the opportunity to identify genes and proteins that may serve as biomarkers in infectious diseases. The identification of biomarkers may enable the development of exciting potential clinical applications in which genes and proteins that are differentially expressed in healthy and infected individuals can be investigated (The International HapMap Project, 2003). These approaches may provide detailed insight into the pathogenesis of disease, host pathogen interactions, and disease-specific expression patterns. In addition, diagnostic and prognostic biomarkers and markers that monitor disease or response to therapy may be developed using these technologies.

This chapter describes recent advances in genomics and proteomics in the field of biomarkers for infectious diseases and summarizes current clinical applications and future perspectives.

2 Molecular Profiling: Current Techniques

2.1 Genomics

Single nucleotide polymorphisms (SNPs), single base pair changes at specific spots in the genome, are the most common type of genetic variation. The human genome carries over 10 million nucleotides that vary in at least 1% of the population. Currently, approximately 6 million nucleotides have been validated and this number is still growing (Bryant et al., 2004; Walker and Siminovitch, 2007). Although SNPs are the changes most frequently explored using high throughput technologies, other genetic variations are also common in the human genome and may influence the individual's susceptibility to disease. These include variations in gene copy number, repeating sequence motifs, insertions, and deletions (Crawford et al., 2005).

The completion of the human genome map in combination with the development of microarray-based comparative genomic hybridization and genome-wide SNP platforms have permitted the screening of the entire genome to identify genetic loci linked to certain diseases, susceptibility to disease or response to therapy (Feuk et al., 2006). These genome wide association studies allow identification of genetic risk factors for a wide variety of common and more complex diseases by measuring hundreds of thousands of genetic variants simultaneously.

Using SNP platforms, MalariaGen (2009), a genomic epidemiological network, and the Wellcome Trust Case-Control Consortium (WTCCC, 2006) have performed a large scale study on disease susceptibility. Both consortia have analyzed up to 500,000 SNPs in thousands of African individuals diagnosed with malaria or tuberculosis between 2006 and 2008. In addition, the WTCCC will include approximately 2,000 cases and 3,000 controls for 8 other diseases, which makes it one of the biggest projects aimed at the identification of genetic variations that may predispose a patient to disease (Genome-wide Association Study, 2007). These genome-wide analyses may contribute to the identification of individuals at risk and produce more effective prevention strategies and individual treatment strategies.

2.2 Transcriptomics

Where genomics provides information on genetic susceptibility to certain infectious diseases, transcriptomics provides information on the activity of genes at a certain moment under certain conditions. It is the study of the complete set of RNA transcripts produced by the genome. During all biological processes, part of the genome is specifically transcribed into messenger RNA (mRNA, transcriptome) and translated to proteins (proteome). The transcriptome can be analyzed using gene expression microarrays. These chips contain either the whole genome or a subset of specific genes. mRNA is extracted from experimental samples, reverse transcribed and labeled with fluorescent dyes. The extracted labeled cDNA is then hybridized with the microarray and the fluorescence of the array is determined using an array scanner. Following image analysis, the data are subjected to bioinformatics processes to identify statistically significant changes in gene expression between different samples. The technique can be used to characterize gene expression in both pathogen and host, providing detailed insight into host-pathogen interactions during infection (Liu et al., 2006). To study them on a molecular level, Kawada et al. generated gene-expression profiles in peripheral blood mononuclear cells (PBMCs) isolated from children with influenza virus infection. Many genes associated with the immune response such as interferon regulated genes appeared to be strongly upregulated during influenza infection. In addition, they compared gene expression profiles of influenza-infected children with and without convulsions. They found that transcription levels of pro-inflammatory cytokine genes in patients with a febrile convulsion were not significantly different from those in patients without febrile convulsion. This kind of approach may help to clarify the pathogenesis of influenza and its neurological complications (Kawada et al., 2006).

2.3 Proteomics

While gene expression profiles may not completely correlate with intracellular protein content, proteomics can provide insight into the structure and dynamics of the end product, proteins. Proteomics is the study of the proteome, the complete set of proteins, their modifications, interactions, and localization. Proteomic technologies enable detailed analyses of protein expression and evaluation of post-translational modification and protein stability and turnover that cannot be assessed by genomic and transcriptomic profiling alone. For many years, two-dimensional gel electrophoresis has been the standard technology to isolate specific proteins and allow protein identification by subsequent mass spectrometry. During the past decade, mass spectrometry has improved and now enables the analysis of protein expression, structure, and function without the need for labor intensive and time consuming electrophoresis (Graves and Haystead, 2002; Patterson and Aebersold, 2003). In addition, several mass spectrometry-based approaches have been developed that allow the relative or absolute quantification of proteins.

In an attempt to identify biomarkers and to develop screening tools with high sensitivity and specificity, proteomics technology has been applied to analyze biofluids such as serum, saliva, or urine. A commonly used technique is a surface enhanced laser desorption/ionization time-of-flight (SELDI-TOF), which is a proteomics technique that allows the identification of large numbers of proteins in a short period of time. This technique provides a specific mass spectral profile from each analyzed protein sample. By comparing profiles from affected individuals to those derived from healthy controls, a specific protein signature is obtained (Coombes et al., 2005; Hodgetts et al., 2007). Agranoff and colleagues have used it to characterise distinct profiles for several microbial infections and to investigate serum responsiveness to *M. tuberculosis* identifying serum biomarkers from patients with advanced tuberculosis. SELDI-TOF has the potential to identify tuberculosis at an early stage, assisting early diagnosis and therapy, which is important for favorable outcome (Agranoff et al., 2005).

3 Clinical Applications in Pediatric Infectious Diseases

The recent advances in proteomic and genomic technologies have allowed the identification of genes and proteins that may serve as biomarkers for the diagnosis and monitoring of infectious diseases. Application of knowledge from these technologies into clinical practice is still at an early stage. In the following sections, we will first discuss current literature on the contribution of the aforementioned technologies for the determination of disease pathogenesis, for the susceptibility to infection, and for diagnostics. In the next section, we will discuss clinical applications of proteomic and transcriptomic technologies, and, in the last section, we will focus on their future perspectives.

3.1 What Has Been Studied up to now?

With increasing use of genomic and proteomic technologies, more insight has been obtained into host-pathogen interactions and pathogenesis of infectious diseases. The response of the host to pathogens is reflected in changes in gene expression and can be measured by microarray based gene expression technologies. Likewise, transcriptional profiling studies have proven to be a powerful approach for analyzing and understanding host–pathogen interactions. Based upon the host response to various pathogens, Jenner et al. have identified common and more specific gene expression patterns. They collected and systematically compared transcriptional profiling datasets from 32 published microarray-based in vitro studies which collectively examined 77 different host-pathogen interactions. In response to this wide variety of pathogens, they identified a cluster of 511 genes that share a common response upon infection with a pathogen. According to the localization of the cell in which they function, these genes have been clustered into different functional groups in order to provide an overview of cellular physiology involved in the common host response. Moreover, analyzing different transcriptional profiling studies also revealed pathogen specific gene expression patterns. Several host genes of the aforementioned common host response were found to be downregulated in the presence of pathogens or specific pathogen proteins. These transcriptional profiling studies has provided insight into how micro-organisms alter host gene expression patterns to subvert the immune responses. For example, transcriptional profiling has identified that viruses such as herpes simplex virus (HSV)-1, human cytomegalovirus, and human papillomavirus-31 are partially or completely able to inhibit the induction of Interferon stimulated genes, which have a central role in the defense against viruses (Jenner and Young, 2005).

Streptococcus pneumoniae and influenza virus are the most common causes of pneumonia. Consequently, they contribute to substantial morbidity and mortality worldwide. It has been known for years that influenza infection predisposes to secondary bacterial infection, mainly caused by *S. pneumoniae* and *S. aureus*. The catastrophic influenza A pandemic in 1918 in which approximately 40–50 million persons were killed, may have involved synergy between influenza and pneumococcal infections. Gene expression profiling is a powerful tool to explore the molecular mechanism of synergy between pathogens. By host gene expression analyses of the lungs in mice, Rousseau et al. differentiated pneumococcal infections from influenza. Rosseau et al., have also identified common gene expression patterns in infectious disease as well as unique pathogen-specific gene expression signatures that may help clarify the mechanisms behind the synergy between influenza virus and *S. pneumoniae* (Rosseau et al., 2007). In response to influenza infection, Tong et al. performed gene expression analyses of middle ear epithelial cells (Tong et al., 2004). Tong et al. suggest that increased expression of inflammatory mediator genes such as IP-10 and CXCL11 could lead to a shift in *S. pneumoniae* adherence by activation of host epithelial and endothelial cells, providing a favorable environment in the middle ear cavity for a secondary bacterial infection with *S. pneumoniae*.

The response upon exposure to pathogens varies widely between individuals. Some people are more susceptible to a certain infection than others. This differential susceptibility is partly caused by genetic variations between individuals that may predispose either to development of disease or to a more severe course. Although the large variation in clinical responses among individuals, also depends upon environmental and microbial factors. The classic example of host genetic susceptibility is the resistance of heterozygous hemoglobin S individuals to malaria infection (caused by *Plasmodium falciparum*) (Allison, 1954). Other approaches used to elucidate genetic and environmental effects on infections include studies in identical and non-identical twins (Hill, 1998; Jepson et al., 2001; Jepson et al., 1995) and comparisons of risk in adopted children and their biological and adoptive parents. One such study suggested that adopted children with a biological parent who died early of an infectious disease had a higher risk of mortality from similar infections while the death of an adoptive parent due to infection had no influence on the risk of disease in the children (Sorensen et al., 1988).

Recent advances in microarray technologies have enabled genome wide searches for genes influencing susceptibility to infectious diseases. Analysis of genetic susceptibility aims to link the genetic code (genotype) to the risk of a certain disease state (phenotype). Given the large number of human genes, many with unknown function, genome wide studies have the advantage that previously unconsidered genes can be identified and provide more sensitivity for the detection of subtle genetic effects and gene recruitment in affected individuals. However, many reported genetic associations have not been replicated in subsequent studies, and, for secure results, large numbers of affected and unaffected individuals are required. Furthermore, because of the complexity of data analysis, microarray technologies are time and labor intensive (Xavier and Rioux, 2008; Cooke and Hill, 2001; Hirschhorn and Daly, 2005). Nevertheless, several large scale population based studies have been performed and support the role of genome wide searches in the identification of genes influencing disease susceptibility (WTCCC, 2006; O'Brien and Nelson, 2004; An et al., 2007; Hill, 2006). For example, a genetic association study performed by O'Brien et al. has led to the identification of various genetic factors that affect HIV-1/AIDS. Genetic association analysis of several large cohorts of HIV infected individuals resulted in the identification of 14 AIDS restriction genes, which are human genes with polymorphic variants that influence the outcome of HIV-1 exposure or infection. The study of O'Brien et al., illustrates the discovery of previously unknown genes involved in susceptibility to infection using SNP haplotype-based association studies in clinically well-described epidemiological cohorts (O'Brien and Nelson, 2004).

Another potential application of microarray technologies is diagnosis of infection both by direct and indirect identification of pathogens. For example, microarrays composed of DNA sequences of various pathogens allow the identification of many organisms in a single test. Wang et al. developed an array composed of all fully sequenced reference viral genomes that allows the detection of approximately 1000 viruses (Wang et al., 2002). Moreover, by sequencing hybridized material of unknown pathogens, this array permits identification of new viruses, and, in 2003, it proved successful in the global effort to identify the novel corona virus associated with severe acute respiratory syndrome (SARS) (Wang et al., 2002; Ksiazek et al., 2003).

In contrast to direct identification, infections can also be characterised indirectly through specific host responses. An advantage of such pathogen-specific molecular signatures in the host is that they may be present at various stages of infection, even when the pathogen is not detectable using standard or direct diagnostic tests. Ramilo et al. (2007), used gene expression analysis to diagnose different pathogen fingerprints in pediatric patients with respiratory infections caused by influenza A virus and Gram-negative (*E.coli*) or Gram-positive (*Staphylococcus aureus* and *Streptococcus pneumoniae*) bacteria. Classifier genes, which discriminate influenza A from bacterial infections (*S. pneumoniae and S. aureus*) and *E. coli* from *S. aureus* infections, were identified and validated. Another example of "–omic" technology use in diagnosing infection is the development of a protein based signature for diagnosing trypanosomiasis or sleeping sickness, which affects half a million people yearly in sub-Saharan Africa. Trypanosomiasis, if left untreated, is a debilitating disease with a lethaloutcome; it was successfully controlled in the past, but, since the 1970s, has re-emerged as an epidemic of immense proportions causing huge, yet widely underestimated morbidity and mortality of up to 50,000 cases every year (Stich et al., 2002). Establishing the diagnosis remains complicated, as current diagnostic tests lack the sensitivity to detect low parasite loads in peripheral blood. Papadopoulos et al. analyzed serum samples from patients and controls using SELDI –TOF mass spectrometry and identified distinct serum proteomic signatures in both groups (Papadopoulos et al., 2004). After depleting serum samples of antibody components, the authors identified two prominent protein peaks at 23/24 and 47 kDa in patients. These proteomic signatures may provide the basis for new diagnostic tests and alternative methods to monitor the host response to treatment. Moreover, additional characterization of these differentially expressed proteins may allow the development of simpler, cheaper antibody based tests (Papadopoulos et al., 2004; Agranoff et al., 2005).

3.2 Current and Potential Clinical Applications of "–omic" Technologies

More than 50% of all children admitted to the hospital have fever or other non-specific symptoms related to infection (Schaad, 1997). Although not necessarily suffering from bacterial infection, a significant proportion of these children will receive antibiotics. Although clinical history, physical examination and conventional diagnostic investigations (e.g., x-rays, blood tests) may point to an extent towards cause, pathogen identification remains difficult or even impossible. During episodes of acute fever, clinicians prefer to rely on cultures taken from the site of infection. However, such cultures often cannot be obtained at the right time or from the relevant site (e.g., middle ear or lungs) and results are not available for at least 24 h after sampling so that pathogens often remain undetected. Furthermore, contamination and colonization, particularly in upper airway samples, can obscure results. Gene expression profiles may identify bacterial pathogens and discriminate between bacterial infections, infections caused by other pathogens, and non-infectious causes of fever (like auto-inflammatory diseases). Using microarrays, organisms can be identified either directly or indirectly through their effects on host gene expression.

Tang et al. showed that gene expression profiling of neutrophils can distinguish sepsis from non-infectious inflammation (e.g., Systematic Inflammatory Response Syndrome, SIRS) in intensive care patients. They performed microarrays on a cohort of septic (N=71) and non-septic (N=23) patients and identified 50 classifier genes differentially expressed between the two groups, which are involved in inflammatory responses, immune regulation, and mitochondrial function. Broadly genes involved in the upregulation of immune responses were expressed less in patients with sepsis than in control patients, whereas genes involved in down regulation were expressed more, suggesting that sepsis may have an inhibitory effect on immune regulation. Pathway analyses support the finding that the immune regulation is inhibited during sepsis by showing that genes involved in the NF-B pathway were expressed less in patients with sepsis, whereas the inhibitory gene *NFKBIA* was expressed more. A prediction model for disease severity was developed from these data and validated in a second more heterogeneous patient group (Tang et al., 2007). A major advantage of gene expression profiling is that it may enable the development of less expensive diagnostic tools such as quantitative real time-polymerase chain reaction (RT-PCR) detection and quantification of specific DNA sequences in septic patients instead of entire gene expression profiles.

Children with auto-inflammatory diseases often present with non-specific systemic symptoms like rash and fever, which precede more specific symptoms like arthritis. Diagnosis of inflammatory diseases is often difficult due to the presentation with non-specific symptoms and the low incidence of these disease. Patients are often empirically treated for more likely causes of symptoms, including infections. Such delay in diagnosis and initiation of appropriate treatment is suboptimal for the child and may also result in misuse of antibiotics, contributing to emergence of antibiotic resistance. To differentiate patients with auto-inflammatory diseases (e.g., systemic onset juvenile idiopathic arthritis) from patients with acute viral and bacterial infections Allantez et al. analyzed leukocyte gene expression profiles of different blood leukocyte subpopulations that were obtained from these patients. Based on their results, a specific blood signature was developed that enabled differentiation between infection and other febrile inflammatory diseases (Allantaz et al., 2007).

Along with permitting characterization of infections when pathogens are not directly detectable, measurement of specific gene expression by the host can potentially permit distinction between colonization and infection with pathogenic micro-organisms. For example, bacterial secondary infections in children with primary viral lower respiratory tract infections are often diagnosed based on cultures from upper airway samples. The question remains whether the detected organism is really the cause of infection or whether it is just a contaminant from the upper airway (Jacobs and Dagan, 2004). The development of a diagnostic test based on gene expression patterns in the host may provide more specific information. In the future, the development of such diagnostic tools may help the clinician choose an effective treatment strategy and reduce inappropriate antibiotic use.

A diagnostic delay in infectious diseases can lead to delayed initiation of therapy, severe complications, and long term consequences that may include death. Such a delay in diagnosis may be prevented by the development of diagnostic biomarkers. Encephalitis, for example, is a complex, severe, neurological syndrome associated with significant morbidity and mortality. It is characterized by inflammation of the brain parenchyma, and children often present with drowsiness, fever, headache, seizures, or focal neurological signs. A delay in treatment may lead to irreversible brain damage. Diagnosis is often presumptive and based on clinical characteristics or increased serological antibody titers. Unfortunately, the causative pathogen is often hard to detect in the central nervous system itself. The final diagnosis is sometimes based on the detection of pathogens in cultures from other sites such as the respiratory tract (Glaser et al., 2006). Indirectly diagnosing encephalitis based on respiratory samples is rather inaccurate and demonstrates the need for new and better diagnostic tools. New microarray and proteomics technologies may contribute to improved diagnosis and treatment. Microarrays have been developed to simultaneously identify different viral and bacterial pathogens in cerebrospinal fluid (CSF) (Boriskin et al., 2004; Ben et al., 2008). To our knowledge, gene expression studies for differentiating pathogens based on the host response have not yet been performed. The differentiation of pathogens based on the host response may provide better insight in pathogenesis and disease specific profiles in blood or cerebrospinal fluid; it may also prove useful in diagnosing encephalitis. The identification of biomarkers related to clinical profiles or recognition of subgroups in encephalitis may help predict outcome and provide insight into the efficacy of therapy.

Another infection in which diagnostic difficulties often lead to a delay in appropriate therapy is infective endocarditis. The clinical diagnosis of infective endocarditis may be difficult, as fever can be the only symptom. Rapid diagnosis followed by appropriate treatment is of critical importance for survival. However, in 3–31% of patients, causative pathogens remain undetected. Fenollar et al. analyzed serum samples from 88 patients with a clinical suspicion of endocarditis. They identified 66 different protein peaks in patients with confirmed endocarditis as compared to those in whom the diagnosis was excluded (Fenollar et al., 2006). From these 66, they developed a diagnostic assay based on 7 protein peaks. Despite this limited number of differentially expressed proteins, the test was still able to classify the majority (88%) of patients correctly.

3.3 Future Perspectives for Biomarker Development

Proteomic and genomic technologies have been shown to contribute to improved insight into disease pathogenesis and may be useful in diagnosing infections and providing information about disease susceptibility. However, clinical application of these technologies has not yet been developed. Future research should focus on the validation of previously identified biomarkers as well as the development of new diagnostic assays.

The relationship between gene expression profiles and disease outcome is another interesting field of research. Prognostic biomarkers could be import in infectious diseases, helping predict likely disease course and a selection of patients most likely to benefit from treatment. At present, their clinical use is limited to the field of cancer research where several studies suggest that molecular classification of tumors, based on gene expression, may identify distinct prognostic subtypes. Alizedah et al. have detected two subtypes of diffuse B-cell lymphoma with different survival patterns (Alizadeh et al., 2000). The Mamma print, a test developed by the Dutch Cancer Institute, identifies different breast cancer subtypes by analysis of expression profiles involving 70 genes indicative of poor prognosis (van't Veer et al., 2002; van't Veer and Bernards, 2008). These studies are based on hierarchical clustering of subgroups with similar gene expression profiles. Hierarchical clustering methods in gene expression analyses might prove useful in pediatric infectious diseases; although to date, few studies have been done. Chaussabel et al. have generated a potentially useful framework for the visualization and functional interpretation of microarray based disease-specific transcriptional signatures (Chaussabel et al., 2008), and, in addition, to the identification of biomarkers for monitoring inflammatory disease activity (in SLE) showed its potential value in the evaluation of disease progression and thus prognosis (Chaussabel et al., 2008).

In our department, we are currently conducting a clinical study to identify classifier genes that predict outcome in children suffering from lower respiratory tract viral infections (VIRGO study). Using microarray analyses of blood leukocytes and respiratory samples, we aim to identify biomarkers that distinguish children with a relatively mild course of disease from those who will deteriorate and require supplemental oxygen or mechanical ventilation. In the early phase of infection, these biomarkers may help decide whether a child needs to be hospitalized.

Another potential application is to guide treatment by allowing therapy to be tailored both to the specific pathogen, including its antimicrobial resistance properties and to host characteristics, including the immune response, which leads to more focused drug use and improved outcome. An early example of genotype-guided, individualized treatment strategy is the adjustment of isoniazid dosing regimen in adults with tuberculosis. N-acetyltransferase type 2 (NAT2) plays an important role in isoniazid metabolism and genetic polymorphisms of this enzyme can alter the response to the drug. Determining the NAT2 genotype prior to isoniazid administration can predict pharmacokinetic variability and therapeutic response (Kinzig-Schippers et al., 2005).

Individualized treatment strategies will be extremely helpful in treating tuberculosis in children. At present, children with tuberculosis are treated with up to four tuberculostatic agents for two months followed by a two drug regimen during a 4 month continuation phase. To date, there is no laboratory tool to monitor response to therapy. Moreover, difficulties in identification of *M. tuberculosis* from, for example, induced sputum or gastric lavage, contribute to diagnostic and therapeutic uncertainty (Newton et al., 2008). Consequently, non-specific clinical features such as symptom improvement, weight gain, and radiological features of chest disease are used as markers for therapeutic response (Donald and Schaff, 2007). The identification of biomarkers for tuberculosis disease activity may provide more specific monitoring strategies leading to more focused prescribing, fewer adverse effects, and less multidrug resistance.

4 Conclusions

Diagnostic uncertainty in infectious disease may lead to delayed diagnosis and inappropriate use of antibiotics. The development of diagnostic biomarkers for infectious diseases may result in more rapid diagnosis, more reliable discrimination between infection and non-infectious diseases, more improved management, better course and outcomes, and less inappropriate use of antibiotics.

Microarrays and proteomic technologies are beginning to contribute to improved understanding of the pathogenesis of a wide variety of infectious diseases and have great prospects for the future. These technologies can be targeted both at direct pathogen detection and at characterization of the host response, which may also assist in diagnosis and disease monitoring as well as predicting the individual's susceptibility to disease, response to medical therapy, and overall prognosis. Although promising, the clinical application of these technologies in infectious diseases is limited at present. Current research focuses on sophisticated highly specialized techniques, but future work will need to be directed at clinical validation studies to collect data on clinical applicability, accuracy and cost effectiveness. Translating biomarker research into clinically useful tests will be difficult and time and labor intensive. The ultimate goal is to develop clinically relevant, cheap, rapid diagnostic and prognostic biomarker tests which use biological samples that are easy to obtain from the patient and which generate reliable and easily interpreted results.

References

Agranoff, D., Stich, A., Abel, P., & Krishna, S. (2005). Proteomic fingerprinting for the diagnosis of human African trypanosomiasis. *Trends Parasitol* 21(4), 154–157.

Alizadeh, A.A., Eisen, M.B., Davis, R.E., Ma, C., Lossos, I.S., Rosenwald, A. et al. (2000). Distinct types of diffuse large B-cell lymphoma identified by gene expression profiling. *Nature* 403(6769), 503–511.

Allantaz, F., Chaussabel, D., Stichweh, D., Bennett, L., Allman, W. et al. (2007). Blood leukocyte microarrays to diagnose systemic onset juvenile idiopathic arthritis and follow the response to IL-1 blockade. *J Exp Med* 204(9), 2131–2144.

Allison, A.C. (1954). Protection afforded by sickle-cell trait against subtertian malareal infection. *Br Med J* 6, 1(4857), 290–294.

An, P., Wang, L.H., Hutcheson-Dilks, H., Nelson, G., Donfield, S., Goedert, J.J., & Rinaldo, C.R. et al. (2007). Regulatory polymorphisms in the cyclophilin A gene, PPIA, accelerate progression to AIDS. *PLoS Pathog* 3(6), e88.

Ben, R.J., Kung, S., Chang, F.Y., Lu, J.J., Feng, N.H., & Hsieh, Y.D. (2008). Rapid diagnosis of bacterial meningitis using a microarray. *J Formos Med Assoc* 107(6), 448–453.

Boriskin, Y.S., Rice, P.S., Stabler, R.A., Hinds, J.A., Ghusein, H., Vass, K. et al. (2004). DNA microarrays for virus detection in cases of central nervous system infection. *J Clin Microbiol* 42(12), 5811–5818.

Bryant, P.A., Venter, D., Robins-Browne, R., & Curtis, N. (2004). Chips with everything: DNA microarrays in infectious diseases. *Lancet Infect Dis* 4(2), 100–111.

Chaussabel, D., Quinn, C., Shen, J., Patel, P., Glaser, C., Baldwin, N. et al. (2008). A modular analysis framework for blood genomics studies: application to systemic lupus erythematosus. *Immunity* 29(1), 150–164.

Cooke, G.S., & Hill, A.V. (2001). Genetics of susceptibility to human infectious disease. *Nat Rev Genet* 2(12), 967–977.

Coombes, K.R., Morris, J.S., Hu, J., Edmonson, S.R., & Baggerly, K.A. (2005). Serum proteomics profiling – a young technology begins to mature. *Nat Biotechnol* 23(3), 291–292.

Crawford, D.C., Akey, D.T., & Nickerson, D.A. (2005). The patterns of natural variation in human genes. *Annu Rev Genomics Hum Genet* (6), 287–313.

Donald, P.R., & Schaaf, H.S. (2007). Old and new drugs for the treatment of tuberculosis in children. *Paediatr Respir Rev* 8(2), 134–141.

Feuk, L., Marshall, C.R., Wintle, R.F., & Scherer, S.W. (2006). Structural variants: changing the landscape of chromosomes and design of disease studies. *Hum Mol Genet* 15(Spec No 1): R57–R66.

Fenollar, F., Goncalves, A., Esterni, B., Azza, S., Habib, G., Borg, J.P., & Raoult, D. (2006). A serum protein signature with high diagnostic value in bacterial endocarditis: results from a study based on surface-enhanced laser desorption/ionization time-of-flight mass spectrometry. *J Infect Dis* 194(10), 1356–1366.

Galetto-Lacour, A., Zamora, S.A., & Gervaix, A. (2003). Bedside procalcitonin and C-reactive protein tests in children with fever without localizing signs of infection seen in a referral center. *Pediatrics* 112(5), 1054–1060.

Genome-wide association study of 14,000 cases of seven common diseases and 3,000 shared controls. (2007). *Nature* 7, 447(7145), 661–678.

Glaser, C.A., Honarmand, S., Anderson, L.J., Schnurr, D.P., Forghani, B., Cossen, C.K. et al. (2006). Beyond viruses: clinical profiles and etiologies associated with encephalitis. *Clin Infect Dis* 43(12), 1565–1577.

Graves, P.R., & Haystead, T.A. (2002). Molecular biologist's guide to proteomics. *Microbiol Mol Biol Rev* 66(1), 39–66.

Herd, D. (2007)In children under age three does procalcitonin help exclude serious bacterial infection in fever without focus? *Arch Dis Child* 92(4), 362–364.

Hill, A.V. (1998). The immunogenetics of human infectious diseases. *Annu Rev Immunol* (16), 593–617.

Hill, A.V. (2006). Aspects of genetic susceptibility to human infectious diseases. *Annu Rev Genet* (40), 469–486.

Hirschhorn, J.N., & Daly, M.J. (2005). Genome-wide association studies for common diseases and complex traits. *Nat Rev Genet* 6(2), 95–108.

Hodgetts, A., Levin, M., Kroll, J.S., & Langford, P.R. (2007). Biomarker discovery in infectious diseases using SELDI. *Future Microbiol* (2), 35–49.

The International HapMap Project. (2003). *Nature* 18, 426(6968), 789–796.

Jacobs, M.R., & Dagan, R. (2004). Antimicrobial resistance among pediatric respiratory tract infections: clinical challenges. *Semin Pediatr Infect Dis* 15(1), 5–20.

Jenner, R.G., & Young, R.A. (2005). Insights into host responses against pathogens from transcriptional profiling. *Nat Rev Microbiol* 3(4), 281–294.

Jepson, A.P., Banya, W.A., Sisay-Joof, F., Hassan-King, M., Bennett, S., & Whittle, H.C. (1995). Genetic regulation of fever in Plasmodium falciparum malaria in Gambian twin children. *J Infect Dis* 172(1), 316–319.

Jepson, A., Fowler, A., Banya, W., Singh, M., Bennett, S., Whittle, H., & Hill, A.V. (2001). Genetic regulation of acquired immune responses to antigens of Mycobacterium tuberculosis: a study of twins in West Africa. *Infect Immun* 69(6), 3989–3994.

Kawada, J., Kimura, H., Kamachi, Y., Nishikawa, K., Taniguchi, M., Nagaoka, K., & Kurahashi, H. et al. (2006). Analysis of gene-expression profiles by oligonucleotide microarray in children with influenza. *J Gen Virol* 87(Pt6), 1677–1683.

Kinzig-Schippers, M., Tomalik-Scharte, D., Jetter, A., Scheidel, B., Jakob, V., Rodamer, M. et al. (2005). Should we use N-acetyltransferase type 2 genotyping to personalize isoniazid doses? *Antimicrob Agents Chemother* 49(5), 1733–1738.

Ksiazek, T.G., Erdman, D., Goldsmith, C.S., Zaki, S.R., Peret, T., Emery, S. et al. (2003). A novel coronavirus associated with severe acute respiratory syndrome. *N Engl J Med* 348(20), 1953–1966.

Liu, M., Popper, S.J., Rubins, K.H., & Relman, D.A. (2006). Early days: genomics and human responses to infection. *Curr Opin Microbiol* 9(3), 312–319.

MalariaGen. (2009). Retrieved from http://malariagen net/access

Newton, S.M., Brent, A.J., Anderson, S., Whittaker, E., & Kampmann, B. (2008). Paediatric tuberculosis. *Lancet Infect Dis* 8(8), 498–510.

O'Brien, S.J., & Nelson, G.W. (2004). Human genes that limit AIDS. *Nat Genet* 36(6), 565–574.

Papadopoulos, M.C., Abel, P.M., Agranoff, D., Stich, A., Tarelli, E., Bell, B.A. et al. (2004). A novel and accurate diagnostic test for human African trypanosomiasis. *Lancet* 363(9418), 1358–1363.

Patterson, S.D., & Aebersold, R.H. (2003). Proteomics: the first decade and beyond. *Nat Genet* 33(Suppl), 311–323.

Ramilo, O., Allman, W., Chung, W., Mejias, A., Ardura, M., Glaser, C. et al. (2007). Gene expression patterns in blood leukocytes discriminate patients with acute infections. *Blood* 109(5), 2066–2077.

Rosseau, S., Hocke, A., Mollenkopf, H., Schmeck, B., Suttorp, N., Kaufmann, S.H., & Zerrahn, J. (2007). Comparative transcriptional profiling of the lung reveals shared and distinct features of Streptococcus pneumoniae and influenza A virus infection. *Immunology* 120(3), 380–391.

Schaad, U.B. (1997). Toward an integrated program for patient care in pediatric infections. *Pediatr Infect Dis J* 16(3 Suppl), S34–S38.

Sorensen, T.I., Nielsen, G.G., Andersen, P.K., & Teasdale, T.W. (1988). Genetic and environmental influences on premature death in adult adoptees. *N Engl J Med* 318(12), 727–732.

Stein, C.E., Inoue, M., & Fat, D.M. (2004). The global mortality of infectious and parasitic diseases in children. *Semin Pediatr Infect Dis* 15(3), 125–129.

Stich, A., Abel, P.M., & Krishna, S. (2002). Human African trypanosomiasis. *BMJ*, 325(7357), 203–206.

Tang, B.M., McLean, A.S., Dawes, I.W., Huang, S.J., & Lin, R.C. (2007). The use of gene-expression profiling to identify candidate genes in human sepsis. *Am J Respir Crit Care Med* 176(7), 676–684.

Tong, H.H., Long, J.P., Li, D., & DeMaria, T.F. (2004). Alteration of gene expression in human middle ear epithelial cells induced by influenza A virus and its implication for the pathogenesis of otitis media. *Microb Pathog* 37(4), 193–204.

van Rossum, A.M., Wulkan, R.W., & Oudesluys-Murphy, A.M. (2004). Procalcitonin as an early marker of infection in neonates and children. *Lancet Infect Dis* 4(10), 620–630.

van't Veer, L.J., & Bernards, R. (2008). Enabling personalized cancer medicine through analysis of gene-expression patterns. *Nature* 452(7187), 564–570.

van't Veer, L.J., Dai, H., van de Vijver, M.J., He, Y.D., Hart, A.A., Mao, M., Peterse, H.L. et al. (2002). Gene expression profiling predicts clinical outcome of breast cancer. *Nature* 415(6871), 530–536.

Walker, E.J., & Siminovitch, K.A. (2007). Primer: genomic and proteomic tools for the molecular dissection of disease. *Nat Clin Pract Rheumatol* 3(10), 580–589.

Wang, D., Coscoy, L., Zylberberg, M., Avila, P.C., Boushey, H.A., Ganem, D., & DeRisi, J.L. (2002). Microarray-based detection and genotyping of viral pathogens. *Proc Natl Acad Sci USA* 26, 99(24), 15687–15692.

Wellcome Trust Case-Control Consortium (WTCCC) (2006). Retrieved from http://ccc.sanger.ac.uk/info/overview.shtml

Xavier, R.J., & Rioux, J.D. (2008). Genome-wide association studies: a new window into immune-mediated diseases. *Nat Rev Immunol* 8(8), 631–643.

The Epidemiology and Management of Non Typhoidal Salmonella Infections

Yamikani Chimalizeni, Kondwani Kawaza, and Elizabeth Molyneux

1 Introduction

Non typhoidal salmonella (NTS) infections affect children all over the world. In well resourced countries they usually present as mild gastro-enteritis, which resolves in a few days without active treatment (Jones et al., 2008; Gordon, 2008). In poorer, often crowded parts of the world, NTS causes serious invasive disease (Diez et al., 2004; Ruiz et al., 2000; Ailal et al., 2004; Preveden et al., 2001; Bahwere et al., 2001; Kariuki et al., 2006a; Enwere et al., 2006; Molyneux et al., 1998; Hill et al., 2007; Berkley et al., 2005; Walsh et al., 2000; Phetsouvanh et al., 2006).

The annual incidence of NTS worldwide is 1.3 billion cases, of whom approximately three million die (O'Ryan et al., 2005). In well resourced parts of the world, where invasive disease is uncommon (less than 5% of all infections), the mortality is less than 2% (Jones et al., 2008; Patrick et al., 2004; Papaevangelou et al., 2004; Galanakis et al., 2007; Weinberger and Keller, 2005). In resource-constrained countries with frequent invasive infections, the mortality is 18–24% (Gordon 2008; Graham et al., 2000a; Nathoo et al., 1996; Lee et al., 2000; Blomberg et al., 2005; Chatterjee et al., 2003; Green and Cheesbrough, 1993).

Over the last 30 years, there has been a worldwide epidemic of NTS, although this now appears to be waning (Patrick et al., 2004; Weinberger and Keller, 2005; Centers for Disease Control and Prevention (CDC), 2000; Centers for Disease Control and Prevention (CDC), 2003; Kariuki et al., 2006b). The prevalence and NTS serovar types are monitored and reported at sentinel sites in several countries (WHO, 2006). While well resourced countries are over-represented, several patterns emerge (Fig. 1).

Unlike typhoid, NTS is found in non-human, as well as human samples. The prevalence in non-human and human reservoirs differs, as does the frequency of different serovars (Fig. 2).

E. Molyneux (✉)
Paediatric Department, College of Medicine, P/Bag 360, Blantyre, Malawi
e-mail: emolyneux@malawi.net

A. Finn et al. (eds.), *Hot Topics in Infection and Immunity in Children VI*, Advances in Experimental Medicine and Biology 659, DOI 10.1007/978-1-4419-0981-7_3,
© Springer Science+Business Media, LLC 2010

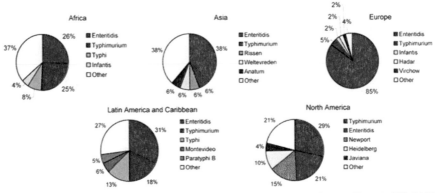

Galanis E et al. EID 2006

Fig. 1 Proportion of most common serotypes reported in humans. Salmonella isolates by world regions 2002. (Galanis et al., 2006)

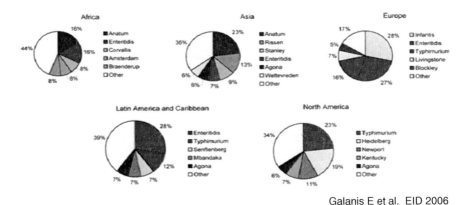

Galanis E et al. EID 2006

Fig. 2 Proportion of most common serotypes reported in animals. Salmonella isolates by world regions 2002. (Galanis et al., 2006)

The epidemic of NTS has been due to *Salmonella enterica* sub sp. *enterica Salmonella Typhimurium* DT 104. It has spread rapidly throughout the world. It is a multidrug-resistant serovar and the pattern of resistance has remained unchanged, globally (Butaye et al., 2006). More recently, it is on the decline, although there has been an increase in the prevalence of *Salmonella enterica* sub sp. *enterica Salmonella* Enteritidis (Gordon, 2008; Graham et al., 2000b).

There are a number of other monitoring systems, augmenting the aforementioned sentinel sites. The CDC has a monitoring network in the United States (Jones et al., 2008). Food Net in the United States now includes reference laboratories from different states. The World Health Organization has established monitoring and training sites on every continent. Eighty four percent of these laboratories report *S. Typhimurium* and *S.* Enteritidis; 50% report *S. infantis* and *S. heidelberg*, and 25%

find cases caused by *S. Newport* (WHO, 2006). These sites over-represent the developed world, where laboratory facilities are available for this monitoring task (WHO, 2006; Helms et al., 2005).

Some serovars are more common in one part of the globe than another; others are geographically limited (Galanis et al., 2006; Jones et al., 2008). For instance, *S.* Enteritidis is the most common serovar in Europe, but in Africa, *S. Typhimurium* is much more common. *S.* Javiana is found only in North America. *S.* Weltevreden is found in seafood and vegetables in South East Asia but is seldom found elsewhere (Wittler and Bass, 1989; Thong et al., 2002).

S. Choleraesuis and *S.* Dublin cause more invasive disease than many other serovars, including *S. Typhimurium* (Jones et al., 2008; Gordon, 2008). In the USA, Jones reports that in the years 1996–2006, *S. Typhimurium* caused 6% of invasive disease, but *S.* Choleraesuis caused 57% and *S.* Dublin, 64% (Jones et al., 2008).

The epidemic of NTS across the world has lead to a rapid rise in multidrug resistance (MDR). Resistance to ampicillin, chloramphenicol, streptomycin, sulphonamides and tetracyclines, (ACSSuT) is now widespread, with sentinel sites reporting 5–80% of all NTS as MDR serovars (Fig. 3).

In Kenya, a rapid rise in the antimicrobial resistance pattern was followed by an equally unaccountable fall in resistance, from 69% in 1994–97 to 11% in 2002–04,

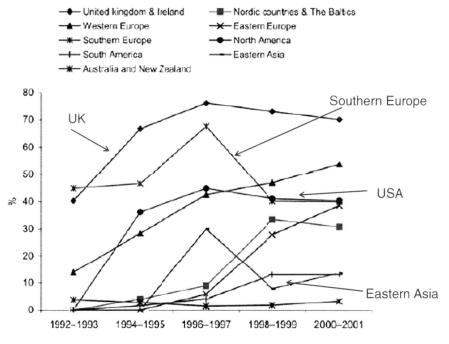

Fig. 3 Multidrug resistant *S. enterica* serovar *S. Typhimurium* as percentage of all *S. Typhimurium* in 9 world regions 1992–2001. (Helms et al., 2005)

despite the continued use of drugs to which the NTS was resistant (Kariuki et al., 2006b). In Malawi, MDR *S*. Enteritidis preceded a similar pattern in *S. Typhimurium* (Fig. 4).

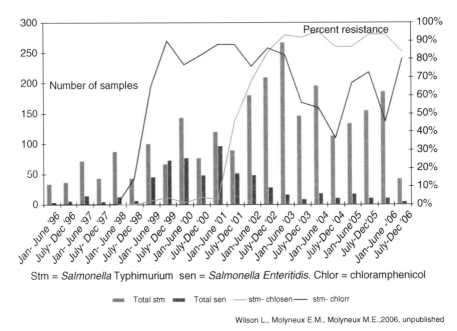

Stm = *Salmonella* Typhimurium sen = *Salmonella Enteritidis*. Chlor = chloramphenicol

▬ Total stm ▬ Total sen ──── stm- chlosen ──── stm- chlorr

Wilson L., Molyneux E.M., Molyneux M.E.,2006, unpublished

Fig. 4 Non Typhoidal salmonella resistance pattern to chloraphenical 1996–2006 in children in Blantyre, Malawi. (Wilson et al., 2006, unpublished)

Resistance appears to be waning here as well. The development of extended-spectrum beta lactam (ESBL) resistance is of great concern. NTS strains that are resistant to quinolones and third-generation cephalosporins have been identified in most parts of the world, including South Africa, Europe, and North and South America (Zaidi et al., 2008; Said et al., 2005) (Fig. 5).

This reduces the availability of suitable, inexpensive antimicrobial therapy in many parts of the world. Despite the dramatic increase in resistance to antimicrobial therapy, the mortality from invasive infections has remained unchanged. It has been suggested that this may be because the more resistant bacteria are less pathogenic (Gordon et al., 2008).

The intestinal form of NTS is an acute ulcerative colitis and does not lead to chronic infection. *S.* Typhi, on the other hand, affects the small bowel, may cause a chronic cholecystitis and 2% of infected people become chronic carriers or "super shedders" (Gordon, 2008). The most common sites of invasive disease are the blood stream, joints, bones, meninges and soft tissues (Preveden et al., 2001; Molyneux et al., 1998; Berkley et al., 2005; Walsh et al., 2000; Galanakis et al., 2007; Graham et al., 2000a; Lee et al., 2000; Green and Cheesbrough, 1993; Yagupsky et al., 2002; Zaidi et al., 1999; Lee et al., 1998; Srifuengfung et al., 2005; Molyneux et al., 2000;

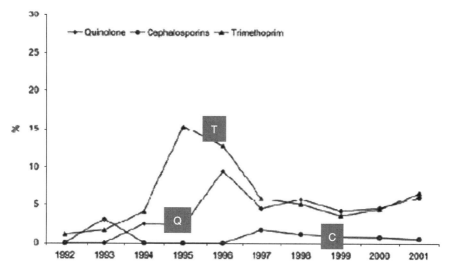

Fig. 5 Proportion of multidrug resistant *S. enterica* serovar *S. Typhimurium* also resistant to quinolones, cephalosporins and trimethoprin 1992–2001. (Helms et al., 2005)

Lavy et al., 1995; Rodríguez et al., 1998). In North America, 2.5% of NTS infections were extra intestinal and many of these were in immunocompromised patients, the very young or the old (Jones et al., 2008). In Spain, the figure was eight percent (Rodríguez et al., 1998) and, in Kuala Lumpur, 5.3% (Lee et al., 1998) of infections are invasive.

In sub-Saharan Africa, NTS has emerged as a common cause of invasive disease (Gordon, 2008; Kariuki et al., 2006a; Molyneux et al., 1998; Hill et al., 2007; Berkley et al., 2005; Walsh et al., 2000; Graham et al., 2000a; Green and Cheesbrough, 1993; Graham et al., 2000b; Oundo et al., 2000; Bachou et al., 2006; Bronzan et al., 2007). In Kenya, in one hospital, 40% of febrile children had NTS in stool samples and 60% of positive blood cultures grew NTS (Noorani et al., 2005). In Malawi 40% of positive blood cultures, 38% of joint cultures and 6% of cases of meningitis grew NTS (Graham et al., 2000a). From the Congo (now the Democratic Republic of Congo), percentages of NTS in blood, joint aspiration and CSF samples were 73, 7, and 5, respectively (Green and Cheesbrough, 1993). Invasive infection occurs notably in the young, (less than 2 years of age), after malarial anemia, and in children who are malnourished (Graham et al., 2000a; Graham et al., 2000b; Bronzan et al., 2007; Noorani et al., 2000; Oundo et al., 2002). NTS is seasonal and is most frequently found at the end of the rainy season, when malarial anemia and malnutrition are also common (Graham et al., 2000b; Maclennan et al., 2008). HIV infection is an underlying factor in NTS invasive disease but it is not as important as anemia and malnutrition (Bachou et al., 2006; Bronzan et al., 2007; Noorani et al., 2005; Oundo et al., 2002; Archibald et al., 2003).

In North America and Europe, the source of NTS infection is foodstuffs. Eggs have been implicated, as have meats (Centers for Disease Control and Prevention

(CDC), 2000; Centers for Disease Control and Prevention (CDC), 2003; McPherson et al., 2006). In Guam, the soil is infected with NTS (Haddock and Nocon, 1986). In the Republic of South Africa, river sediment and water have been found to be reservoirs of NTS pathogenic to humans (Said et al., 2005). In Java, 48% of river water samples were contaminated with NTS (Gracey et al., 1979). In other parts of the world there is more human-to-human spread (Kariuki et al., 2006c; Vaagland et al., 2004; Gracey et al., 1973). Aboriginal children in Australia, Uruguayan children and Kenyans with NTS invasive disease all carry NTS in their oropharynx (Gracey et al., 1973; Hormaeche and Peluffo, 1941; Kariuki et al., 2002). In Kenya and in Mexico, asymptomatic children are carriers of NTS (Kariuki et al., 2006c; Zaidi et al., 2008).

Invasive NTS disease may be focal – with abscess formation in bone, joint infections or meningitis (Srifuengfung et al., 2005; Molyneux et al., 2000; Lavy et al., 1995); however, it very often presents as a fever with no focal lesions. In Malawi, 32% of children with NTS bacteremia were admitted with a diagnosis of malaria; 5% with pneumonia; 11% with gastro-enteritis; 10% with anemia and 10% with sepsis (Graham, 2002). In Gambia, 50% of the cases of NTS septicemia and, in Kenya, 46.5% of cases fitted the WHO criteria for pneumonia (O'Dempsey et al., 1994; Brent et al., 2006).

2 Seasonality of NTS infections

Figure 6 shows how NTS infections are more common in the malaria season and peak towards the end of, and just after the rains have finished. This is also the "hungry season," when the old harvest has finished and the new harvest is

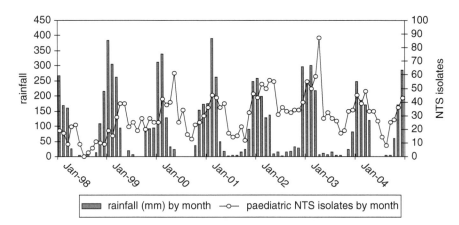

Fig. 6 The relationship between paediatric NTS and rainfall in Blantyre, Malawi 1998–2004. (Gordon et al., 2008)

awaited (Gordon, 2008; Maclennan et al., 2008). In Denmark, NTS causes a summer outbreak of diarrhea and vomiting (Gradel et al., 2007).

3 Malaria, Anemia and NTS infection

The association between malaria and NTS infections in children has been known for a long time. Duggan and Beyer reported the striking number of NTS infections in patients with malaria and especially those with malarial anemia (Duggan and Beyer, 1975). In 1972, Greenwood described the inhibition of *Salmonella* "O" antigen in malaria (Greenwood et al., 1972). In Blantyre, Malawi, of the blood cultures positive in children with malaria, 36% were due to NTS, compared with 5% of *S*. Typhi and 11–29% of other infections (Graham et al., 2000a). This discrepancy is noted in other countries with endemic malaria (Walsh et al., 2000; Oundo et al., 2002).

Why is this so? Malaria causes haemolysis and reduced erythropoesis with an increase in circulating immune complexes and decreased complement, leading to inhibition of Fc receptor mediated phagocytosis (Graham et al., 2000a; Greenwood et al., 1978). Increased erythrophagocytosis of parasitized and non-parasitized red cells alters macrophage and splenic function. Furthermore, unbound iron, released from haemolysed red cells, favours NTS proliferation (Graham et al., 2000a).

NTS infections peak toward the end of the rainy season, which is when the number of cases of malaria decline and anemia caused by malaria is greatest. NTS has been reported to cause 78% of all post-transfusion bacteremia. It is only found in 1% of cases of cerebral malaria but 7% of malarial anemia cases (Graham et al., 2000b), and is especially common in severe anemia (Haematocrit <15%) (Table 1).

Table 1   Anemia and NTS in Blantyre, Malawi. (Graham et al., 2000b; Bronzan et al., 2007)

Hct	<25%	<15%
Non typhoidal *salmonellae*	66%	30%
Salmonella typhi	40%	4%
Gram negative bacteria other than *salmonellae*	39%	4%
Streptococcus pneumoniae	28%	5%

Hct = haematocrit

Nine (31%) of 29 cases of NTS meningitis in Malawian children followed a blood transfusion, compared with 10 (1.8%) of 568 cases who developed meningitis from other causes (Graham et al., 2000c).

4 Malnutrition and NTS Infection

From South Africa in the 1980s, Berkowitz described a higher mortality of NTS with malnutrition (38.7%), compared with 14.5% in the well nourished (Berkowitz, 1992). In 1983, Lloyd Evans et al. in Nigeria, reported that 28.7% of malnourished children compared with 5.3% of normally nourished children excreted NTS over several weeks (Lloyd Evans et al., 1983). Norton et al. in a hospital study from Lilongwe, Malawi, found that malnutrition and not HIV infection was associated with NTS bacteremia (Norton et al., 2004). In a report of bacteremia in malnourished Ugandan children, those with severe malnutrition were more likely to have a bacteremia, and the cause was more likely to be a NTS (OR 5.4, CI 1.6–17.4) (Bachou et al., 2006). In another study from Mulago, Uganda, 22% of 134 malnourished children aged 6–59 months, 44% of whom were HIV infected, had a bacteremia. More than 70% were caused by Gram-negative bacteria, of which 67% were either salmonellae or *E. coli*. Hypoglycemia was a strong predictor of bacteremia. p 0.007 (Babirekere–Iriso et al., 2006)

5 HIV and NTS infection

In adults, the association of HIV infection and NTS is well recognized (Gordon, 2008; Gordon et al., 2002). An HIV-infected adult with a CD4 count of less than 200/ml is 48 times more likely to have invasive NTS disease than an uninfected adult (Gordon, 2008). Of 68 blood isolates, 13.2% of 168 HIV-positive children grew NTS, compared with 3.6% of 28 blood isolates of 161 HIV negative controls (Nathoo et al., 1996). Several studies have reported the association of advanced HIV disease with NTS infection in children (Bachou et al., 2006; Bronzan et al., 2007; Chierakul et al., 2004).

6 NTS Recurrence and Relapse

Nearly half (19/44; 43%) of adult HIV-infected patients had a first recurrence of NTS infection within 3–186 days. Five of 19 had multiple recurrences (Gordon, 2008). Of 61 Malawian children with NTS meningitis, two recurred or relapsed. Relapses were almost unknown in other HIV-uninfected children (Molyneux et al., 2000).

7 Treatment of NTS Infections

Simple febrile gastro-enteritis caused by NTS in the otherwise well nourished and immune-efficient child requires no antimicrobial therapy. Antibiotics may prolong the illness and the excretion of bacteria (Jones et al., 2008).

It is notably difficult to clear salmonella CNS infections, and prolonged treatment is recommended to try to overcome the problem (American Academy of Pediatrics, Committee on Infectious Diseases, 2000; Price et al., 2000). The American Academy of Pediatrics recommends four weeks of antimicrobial therapy for meningitis, and still records a significant relapse. Complications are common even in immunocompetent children (American Academy of Pediatrics Committee on Infectious Diseases, 2000; Price et al., 2000). In immunodeficient patients, recurrence is common and the serovar is the same for each infectious episode (Gordon, 2008).

Treatment depends on the focus of infection, the age of the child, the severity of the illness, the presence of underlying immunosuppression and the local pattern of antimicrobial resistance. In invasive disease where most NTS are known to be resistant to chloramphenicol, and especially in meningitis, it is safer to initiate treatment with an effective antibiotic such as a third-generation cephalosporin (Kinsella et al., 1987). The AAP recommends four weeks of antibiotic treatment. Ceftriaxone (substituted with cefotaxime in neonates) and ciprofloxacin are used in combination therapy (Eliopoulos and Eliopoulos, 1989). Both drugs have good blood-brain barrier and intracellular penetration (Price et al., 2000; Eliopoulos and Eliopoulos, 1989; Hampel et al., 1997; Jick, 1997; Green et al., 1993; Scheld, 1989). Ciprofloxacin achieves rather poor CSF drug levels and is best not used in isolation. Meropenem, imopenem and, more recently, trovofloxacin have been used with excellent results (Price et al., 2000; Koc et al., 1997; Sáez-Llorens et al., 2002).

Joint infections often involve the shoulder. This occurs especially in toddlers who are carried on their mothers' backs and are swung into position by the mother grasping the upper arm and pivoting the child's body weight on the shoulder joint. Minor local trauma may lead to seeding of NTS in the presence of a low-grade bacteremia (Molyneux and French, 1982). Joint infections require antibiotic therapy and surgical intervention. The intervention depends on the specialty of the physician caring for the child and may require arthroscopic washouts, or aspiration and/or incision and drainage, followed by about four weeks of antibiotics. Most clinicians give intravenous antibiotics until the systemic symptoms settle and then change, or oral therapy for the rest of the course (Lavy, 2007).

Bacteremia requires a seven-day course of antibiotics, but this may need to be prolonged in children with immunosuppression.

Ciprofloxacin has been given prophylactically to HIV-infected patients who have frequent relapses of NTS bacteremia, but with antiretroviral treatment available, this should be commenced as soon as possible.

An effective vaccine would prevent many infections and deaths. The oral typhoid vaccines have shown some coincidental protection from NTS infections, but a more directed vaccine is required (Levine et al., 2007; Maclennan et al., 2008).

References

Ailal, F., Bousfiha, A.A., Jouhadi, Z., Adnane, F., & Abid, A. (2004). Forty-one pediatric cases of non-typhoidal salmonellosis. *Med Mal Infect* 34(5), 206–291.

American Academy of Pediatrics Committee on Infectious Diseases. (2000). In G. Peter (Ed.) *Salmonella* Infections. *Report of the Committee on Infectious Diseases*. (25th ed.). p. 503. Elk Grove Village, IL: American Academy of Pediatrics.

Archibald, L.K., Kazembe, P.N., Nwanyanwu, O., Mwansambo, C., Reller, L.B., & Jarvis, W.R. (2003). Epidemiology of bloodstream infections in a bacille Calmette-Guérin-vaccinated pediatric population in Malawi. *J Infect Dis* 188(2), 202–208.

Babirekere-Iriso, E., Musoke, P., & Kekitiinwa, A. (2006). Bacteremia in severely malnourished children in an HIV-endemic setting. *Ann Trop Paediatr* 26(4), 319–328.

Bachou, H., Tylleskär, T., Kaddu-Mulindwa, D.H., & Tumwine, J.K. (2006). Bacteremia among severely malnourished children infected and uninfected with the human immunodeficiency virus-1 in Kampala, Uganda. *BMC Infect Dis* (6), 160.

Bahwere, P., Levy, J., Hennart, P., Donnen, P., Lomoyo, W., Dramaix-Wilmet, M., Butzler, J.P., & De Mol, P. (2001). Community-acquired bacteremia among hospitalized children in rural central Africa. *Int J Infect Dis* 5(4), 180–188.

Berkley, J.A., Lowe, B.S., Mwangi, I., Williams, T., Bauni, E., Mwarumba, S., Ngetsa, C., Slack, M.P., Njenga, S., Hart, C.A., Maitland, K., English, M., Marsh, K., & Scott, J.A. (2005). Bacteremia among children admitted to a rural hospital in Kenya. *N Engl J Med* 352(1), 39–47.

Berkowitz, F.E. (1992). Infections in children with severe protein energy malnutrition. *PIDJ* (11), 750–759.

Blomberg, B., Jureen, R., Manji, K.P., Tamim, B.S., Mwakagile, D.S., Urassa, W.K., Fataki, M., Msangi, V., Tellevik, M.G., Maselle, S.Y., & Langeland, N. (2005). High rate of fatal cases of paediatric septicaemia caused by gram-negative bacteria with extended-spectrum beta-lactamases in Dar es Salaam, Tanzania. *J Clin Microbiol* 43(2), 745–749.

Brent, A.J., Oundo, J.O., Mwangi, I., Ochola, L., Lowe, B., & Berkley, J.A. (2006). Salmonella bacteremia in Kenyan children. *Pediatr Infect Dis J* 25(3), 230–236.

Bronzan, R.N., Taylor, T.E., Mwenechanya, J., Tembo, M., Kayira, K., Bwanaisa, L., Njobvu, A., Kondowe, W., Chalira, C., Walsh, A.L., Phiri, A., Wilson, L.K., Molyneux, M.E., & Graham, S.M. (2007). Bacteremia in Malawian children with severe malaria: prevalence, etiology, HIV coinfection, and outcome. *J Infect Dis* 195(6), 895–904.

Butaye, P., Michael, G.B., Schwarz, S., Barrett, T.J., Brisabois, A., & White, D.G. (2006). The clonal spread of multidrug-resistant non-typhi Salmonella serotypes. *Microbes Infect* 8(7), 1891–1897.

Centers for Disease Control and Prevention (CDC) (2000). Outbreaks of Salmonella serotype enteritidis infection associated with eating raw or undercooked shell eggs – United States, 1996–1998. *MMWR Morb Mortal Wkly Rep* 49(4), 73–79.

Centers for Disease Control and Prevention (CDC) (2003). Outbreaks of Salmonella serotype enteritidis infection associated with eating shell eggs – United States, 1999–2001. *MMWR Morb Mortal Wkly Rep* 51(51–52), 1149–1152.

Chatterjee, H., Pai, D., Jagdish, S., Satish, N., Jayadev, D., & Srikanthreddy, P. (2003). Pattern of nontyphoid ileal perforation over three decades in Pondicherry. *Trop Gastroenterol* 24(3), 144–147.

Chierakul, W., Rajanuwong, A., Wuthiekanun, V., Teerawattanasook, N., Gasiprong, M., Simpson, A., Chaowagul, W., & White, N.J. (2004). The changing pattern of bloodstream infections associated with the rise in HIV prevalence in northeastern Thailand. *Trans R Soc Trop Med Hyg* Nov 98(11), 678–686.

Díez Dorado, R., Tagarro García, A., Baquero-Artigao, F., García-Miguel, M.J., Uría González, M.J., Peña García, P., & del Castillo Martín, F. (2004). Non-typhi Salmonella bacteremia in children: an 11-year review *An Pediatr (Barc)* 60(4), 344–348.

Duggan, M.B., & Beyer, L. (1975). Enteric fever in young Yoruba children. *Arch Dis Child* (50), 67–71

Eliopoulos, G.M., & Eliopoulos, C.T. (1989). Ciprofloxacin in combination with other antimicrobials. *Am J Med* 87, (Suppl 5A), 17S–22S. [Medline]

Enwere, G., Biney, E., Cheung, Y.B., Zaman, S.M., Okoko, B., Oluwalana, C., Vaughan, A., Greenwood, B., Adegbola, R., & Cutts, F.T. (2006). Epidemiologic and clinical characteristics of community-acquired invasive bacterial infections in children aged 2–29 months in The Gambia. *Pediatr Infect Dis J* 25(8), 700–705.

Galanakis, E., Bitsori, M., Maraki, S., Giannakopoulou, C., Samonis, G., & Tselentis, Y. (2007). Invasive non-typhoidal salmonellosis in immunocompetent infants and children. *Int J Infect Dis* 11(1), 36–39.

Galanis, E., Lofo Wong, D.M.,Patrick, M.E., Binzstein, N., Cieslik, A., Chalamchikit, T., Aidara-Kare, A., Ellis, A., Angulo, F.J., & Wegener, H.C. (2006). World Health Organisation Global Salmonella Surveillance: web-based surveillance and global Salmonella distribution, 2000–2002. *EID* 12(3), 381–388

Gordon, M.A., Banda, H.T., Gondwe, M., Gordon, S.B., Boeree, M.J., Walsh, A.L., Corkill, J.E., Hart, C.A., Gilks, C.F., & Molyneux M.E. (2002). Non-typhoidal salmonella bacteremia among HIV-infected Malawian adults: high mortality and frequent recrudescence. *AIDS* 16(12), 1633–1641.

Gordon, M. (2008). Salmonella infections in immunocompromised adults. *JID* 56(6), 413–422.

Gordon, M.A., Graham, S.M., Walsh, A.L., Wilson, L., Phiri, A., Molyneux, E., Zijlstra, E., Heyderman, R., Hart, C.A., & Molyneux, M.E. (2008). Epidemics of Invasive *Salmonella enteritidis* serovar *enteritidits* and *S. Typhimurium* Infections in Adults and Children, associated with Multidrug Resistance in Malawi. *CID*, 46(7), 963–969.

Gracey, M., Ostergaard, P., & Beaman, J. (1973). Oropharyngeal microflora in aboriginal and non-aboriginal children in tropical Australian children: an indicator of environmental contamination. *Aust Paediatr J* (9), 260–322.

Gracey, M., Ostergaard, P., Adnan, S.W., & Iveson, J.B. (1979). Faecal pollution of surface waters in Java. *Trans R Soc Trop Med Hyg* (73), 306–308.

Gradel, K.O., Dethlefsen, C.C., Schønheyder, H., Ejlertsen, T., Sørensen, H., Thomsen, R.W., & Nielsen, H. (2007). Severity of infection and seasonal variation of non-typhoid Salmonella occurrence in humans. *Epidemiol Infect* Jan 135(1), 93–99.

Graham, S.M., Walsh, A.L., Molyneux, E.M., Phiri, A.J., & Molyneux, M.E. (2000a). Clinical presentation of non-typhoidal Salmonella bacteremia in Malawian children. *Trans R Soc Trop Med Hyg* 94(3), 310–314.

Graham, S.M., Molyneux, E.M., Walsh, A.L., Cheeseborough, J.S., Molyneux, M.E., & Hart, C.A. (2000b). Nontyphoidal *Salmonella* infections of children in tropical Africa. *PIDJ*, (19), 1189–1196.

Graham, S.M., Hart, C.A., Molyneux, E.M., Walsh, A.L., & Molyneux, M.E. (2000c). Malaria and Salmonella infections: cause or coincidence? *Trans R Soc Trop Med Hyg* (94), 227.

Graham, S.M. (2002). Salmonellosis is children in developing and developed countries and populations. *Curr Opinion* 15(8), 507–512.

Green, S.D., & Cheesbrough, J.S. (1993). Salmonella bacteremia among young children at a rural hospital in western Zaire. *Ann Trop Paediatr* 13(1), 45–53.

Green, S.D., Ilunga, F., Cheesbrough, J.S., Tillotson, G.S., Hichens, M., & Felmingham, D. (1993). The treatment of neonatal meningitis due to Gram-negative bacilli with ciprofloxacin: evidence of satisfactory penetration into the cerebrospinal fluid. *J Infect* (26), 253–256. [ISI][Medline]

Greenwood, B.M., Bradley–Moore, A.M., Palit, A., & Bryceson, A.D.M. (1972). Immunosuppression in children with malaria. *Lancet* (1), 169–172.

Greenwood, B.M., Stratton, D., Williamson, W.A., & Mohammed, I. (1978). A study of the role of immunological factors in the pathogenesis of the anemia of acute malaria. *Trans R Soc Trop Med Hyg* (72), 378–384.

Haddock, R.L., & Nocon, F. (1986). Salmonella contamination of soil in children's play areas in Guam. *J Environ Health* (49), 158–160

Hampel, B., Hullmann, R., & Schmidt, H. 1997. Ciprofloxacin in paediatrics: worldwide clinical experience based on compassionate use – safety report. *Pediatr Infect Dis J* (16), 127–129. [ISI][Medline]

Helms, M., Ethelberg, S., Melbak, K., & the DT104 Study Group(2005). International *Salmonella* type DT 104 Infections 1992–2001. *EID*, 11(6), 859–867.

Hill, P.C., Onyeama, C.O., Ikumapayi, U.N., Secka, O., Ameyaw, S., Simmonds, N., Donkor, S.A., Howie, S.R., Tapgun, M., Corrah, T., & Adegbola, R.A. (2007). Bacteremia in patients admitted to an urban hospital in West Africa. *BMC Infect Dis*. Jan 26 (7), 2.

Hormaeche, E., & Peluffo, C.A. (1941). Salmonelliasis in infancy and its diagnosis. *PRJ Public Health Trop Med* (17), 99–123.

Jick, S. (1997). Ciprofloxacin safety in a pediatric population. *Pediatr Infect Dis J* (16), 130–134. [ISI][Medline]

Jones, T.F., Ingram, L.A., Cieslak, P.R., Vugia, D.J., Tobin-D'Angelo, M., Hurd, S., Medus, C., Cronquist, A., & Angulo, F.J. (2008). Salmonellosis outcomes differ substantially by Serotype. *JID* (198), 1–6

Kariuki, S., Revathi, G., Gakuya, F., Yamo, V., Muyodi, J., & Hart, C.A. (2002). Lack of clonal relationship between non-typhi Salmonella strain types from humans and those isolated from animals living in close contact. *FEMS Immunol Med Microbiol* 33(3), 165–171.

Kariuki, S., Revathi, G., Kariuki, N., Kiiru, J., Mwituria, J., & Hart, C.A. (2006a). Characterisation of community acquired non-typhoidal Salmonella from bacteremia and diarrhoeal infections in children admitted to hospital in Nairobi, Kenya. *BMC Microbiol. Dec 15*, (6), 101.

Kariuki, S., Revathi, G., Kiiru, J., Lowe, B., Berkley, J.A., & Hart, C.A. (2006b). Decreasing prevalence of antimicrobial resistance in non-typhoidal Salmonella isolated from children with bacteremia in a rural district hospital, Kenya. *Int J Antimicrob Agents* 28(3), 166–171.

Kariuki, S., Revathi, G., Kariuki, N., Kiiru, J., Mwituria, J., Muyodi, J., Githinji, J.W., Kagendo, D., Munyalo, A., & Hart, C.A. (2006c). Invasive multidrug-resistant non-typhoidal Salmonella infections in Africa: zoonotic or anthroponotic transmission? *J Med Microbiol* (55), 585–591.

Kinsella, T.R., Yogev, R., Shulman, S.T., Gilmore, R., & Chadwick, E.G. (1987). Treatment of salmonella meningitis and brain abscess with the new cephalosporins: two case reports and a review of the literature. *Pediatric Infectious Disease Journal* (6), 467–480.

Koc, E., Turkyilmaz, C., Atalay, Y., & Sen, E. (1997). Imipenem for treatment of relapsing *Salmonella* meningitis in a newborn infant. *Acta Paediatr Jpn* (39), 624–625.

Lavy, C.B., Lavy, V.R., & Anderson, I. (1995). Salmonella septic arthritis in Zambian children. *Trop Doct* 25(4), 163–166.

Lavy, C.B. (2007). Septic arthritis in Western and SubSaharan African children - a review. *Int Orthop* 31(2), 137–144.

Lee, W.S., Puthucheary, S.D., & Boey, C.C. (1998). Non-typhoid Salmonella gastroenteritis. *J Paediatr Child Health* 34(4), 387–390.

Lee, W.S., Puthucheary, S.D., & Parasakthi, N. (2000). Extra-intestinal non-typhoidal Salmonella infections in children. *Ann Trop Paediatr* 20(2), 125–129.

Levine, M.M., Ferreccio, C., Black, R.E., Lagos, R., San Martin, O., & Blackwelder, W.C. (2007). Ty21a live oral typhoid vaccine and prevention of paratyphoid fever caused by Salmonella enterica Serovar Paratyphi B. *Clin Infect Dis* 45(Suppl 1), S24–28.

Lloyd Evans, N., Drasar, B.S., & Tomkins, A.M. (1983). A comparison of Campylobacter, Shigellae and Salmonellae in the faeces of malnourished and well-nourished children in the Gambia and northern Nigeria. *Trans R Soc Trop Med Hyg* (77), 245–277.

Maclennan, C.A., Gondwe, E.N., Msefula, C.L., Kingsley, R.A., Thomson, N.R., White, S.A., Goodall, M., Pickard, D.J., Graham, S.M., Dougan, G., Hart, C.A., Molyneux, M.E., & Drayson, M.T. (2008). The neglected role of antibody in protection against bacteremia caused by nontyphoidal strains of Salmonella in African children. *JCI* 118(4), 1553–1562.

McPherson, M.E., Fielding, J.E., Telfer, B., Stephens, N., Combs, B.G., Rice, B.A., Fitzsimmons, G.J., & Gregory, J.E. (2006). A multi-jurisdiction outbreak of *Salmonella Typhimurium* phage type 135 associated with purchasing chicken meat from a supermarket chain. *Commun Dis Intell* 30(4), 449–455.

Molyneux, E.M., & French, G. (1982). Salmonella joint infections in Malawian Children. *J Infect* (4), 131–138.

Molyneux, E., Walsh, A., Phiri, A., & Molyneux, M. (1998). Acute bacterial meningitis in children admitted to the Queen Elizabeth Central Hospital, Blantyre, Malawi in 1996–97 *Trop Med Int Health* 3(8), 610–618.

Molyneux, E.M., Walsh, A.L., Malenga, G., Rogerson, S., & Molyneux, M.E. (2000). Salmonella meningitis in children in Blantyre, Malawi, 1996–1999. *Ann Trop Paediatr* 20(1), 41–44.

Nathoo, K.J., Chigonde, S., Nhembe, M., Ali, M.H., & Mason, P.R. (1996). Community-acquired bacteremia in human immunodeficiency virus-infected children in Harare, Zimbabwe. *Pediatr Infect Dis J* 15(12), 1092–1097.

Noorani, N., Macharia, W.M., Oyatsi, D., & Revathi, G. (2005). Bacterial isolates in severely malnourished children at Kenyatta National Hospital, Nairobi *East Afr Med J* 82(7), 343–348.

Norton, E.B., Archibald, L.K., Nwanyanwu, O.C., Kazembe, P.N., Dobbie, H., Reller, L.B., Jarvis, W.R., & Jason, J. (2004). Clinical predictors of bloodstream infections and mortality in hospitalized Malawian children. *PIDJ* 23(2), 151–155.

O'Dempsey, T.J., McArdle, T.F., Lloyd-Evans, N., et al. (1994). Importance of enteric bacteria as a cause of pneumonia, meningitis and septiciaemia among children in a rural community in the Gambia, West Africa. *PIDJ* (13), 122–128

O'Ryan, M., Prado, V., and Pickering, L.K. (2005). A millennium update on pediatric diarrheal illness in the developing world. *Semin Pediatr Infect Dis* 16(2), 125–136.

Oundo, J.O., Kariuki, S., Maghenda, J.K., & Lowe, B.S. (2000). Antibiotic susceptibility and geno-types of non-typhi Salmonella isolates from children in Kilifi on the Kenya coast. *Trans R Soc Trop Med Hyg* (942), 212–215.

Oundo, J.O., Muli, F., Kariuki, S., Waiyaki, P.G., Iijima, Y., Berkley, J., Kokwaro, G.O., Ngetsa, C.J., Mwarumba, S., Torto, R., & Lowe, B. (2002). Non-typhi salmonella in children with severe malaria. *East Afr Med J* 79(12), 633–639.

Papaevangelou, V., Syriopoulou, V., Charissiadou, A., Pangalis, A., Mostrou, G., & Theodoridou, M. (2004). Salmonella bacteremia in a tertiary children's hospital. *Scand J Infect Dis* 36(8), 547–551.

Patrick, M.E., Adcock, P.M., Gomez, T.M., Altekruse, S.F., Holland, B.H., Tauxe, R.V., & Swerdlow, D.L. (2004). Salmonella enteritidis infections, United States, 1985–1999. *Emerg Infect Dis* 10(1), 1–7.

Phetsouvanh, R., Phongmany, S., Soukaloun, D., Rasachak, B., Soukhaseum, V., Soukhaseum, S., Frichithavong, K., Khounnorath, S., Pengdee, B., Phiasakha, K., Chu, V., Luangxay, K., Rattanavong, S., Sisouk, K., Keolouangkot, V., Mayxay, M., Ramsay, A., Blacksell, S.D., Campbell, J., Martinez-Aussel, B., Heuanvongsy, M., Bounxouei, B., Thammavong, C., Syhavong, B., Strobel, M., Peacock, S.J., White, N.J., & Newton, P.N. (2006). Causes of community-acquired bacteremia and patterns of antimicrobial resistance in Vientiane, Laos. *Am J Trop Med Hyg* 75(5), 978–985.

Preveden, T., Knezević, K., Brkić, S., & Jelesić, Z. (2001). Salmonella bacteremia. *Med Pregl* 54(7–8), 367–370.

Price, E.H., de Louvois, J., & Rella, M. (2000). Workman antibiotics for *Salmonella* meningitis in children. *J Antimicrob Chemother* (46), 653–655.

Rodríguez, M., de Diego, I., & Mendoza, M.C. (1998). Extra-intestinal salmonellosis in a general hospital (1991 to 1996): relationships between Salmonella genomic groups and clinical presentations. *J Clin Microbiol* 36(11), 3291–3296.

Ruiz, M., Rodríguez, J.C., Elía, M., & Royo, G. (2000). Extra-intestinal infections caused by non-typhi Salmonella serotypes. 9 yrs' experience. *Enferm Infecc Microbiol Clin* 18(5), 219–222

Sáez-Llorens, X., McCoig, C., Feris, J.M., Vargas, S.L., Klugman, K.P., Hussey, G.D., Frenck, R.W., Falleiros-Carvalho, L.H., Arguedas, A.G., Bradley, J., Arrieta, A.C., Wald, E.R., Pancorbo, S., McCracken, G.H., Jr, & Marques, S.R. (2002). Trovan menigitis Study Group Quinolone treatment for pediatric bacterial meningitis: a comparative study of trovafloxacin and ceftriaxone with or without vancomycin. *Pediatr Infect Dis J* 21(1), 14–22.

Said, M., le Roux, W.J., Burke, L., Said, H., Paulsen, L., Venter, S.N., Potgeiter, N., Masaobi, D., & de Wit, C.M.E. (2005). Origin, fate and clinical relevance of water-borne pathogens present in surface water. *Water Research Commission Report No. 1398/1/05* Pub August 2005.

Scheld, W.M. (1989). Quinolone therapy for infections of the central nervous system. *Rev Infect Dis* 11(Suppl 5), S1194–1202.

Srifuengfung, S., Chokephaibulkit, K., Yungyuen, T., & Tribuddharat, C. (2005). Salmonella meningitis and antimicrobial susceptibilities. *Southeast Asian J Trop Med Public Health* 36(2), 312–316.

Thong, K.L., Goh, Y.L., Radu, S., Noorzaleha, S., Yasin, R., Koh, Y.T., et al. (2002). Genetic diversity of clinical and environmental strains of *Salmonella enterica* serotype Weltervreden isolated in Malaysia. *J Clin Microbiol* (40), 2498–2503.

Vaagland, H., Blomberg, B., Krüger, C., Naman, N., Jureen, R., & Langeland, N. (2004). Nosocomial outbreak of neonatal Salmonella enterica serotype Enteritidis meningitis in a rural hospital in northern Tanzania. *BMC Infect Dis* (4), 35

Walsh, A.L., Phiri, A.J., Graham, S.M., Molyneux, E.M., & Molyneux, M.E. (2000). Bacteremia in febrile Malawian children: clinical and microbiologic features. *PIDJ* 19(4), 312–318.

Weinberger, M., & Keller, N. (2005). Recent trends in the epidemiology of non-typhoid Salmonella and antimicrobial resistance: the Israeli experience and worldwide review. *Curr Opin Infect Dis* 18(6), 513–521.

Wilson, L., Molyneux, E.M., & Molyneux, M.E. (2006). Unpublished data from the Malawi-Liverpool-Wellcome laboratories and Paediatric Department of the Queen Elizabeth Central Hospital, Blantyre, Malawi.

Wittler, R.R., & Bass, J.W. (1989). Nontyphoidal Salmonella enteric infections and bacteremia. *Pediatr Infect Dis J* 8(6), 364–367.

World Health Organisation. (2006). Salmonella-Surveillance Progress Report (2000–2005) Geneva.

Yagupsky, P., Maimon, N., & Dagan, R. (2002). Increasing incidence of nontyphi Salmonella bacteremia among children living in southern Israel. *Int J Infect Dis* 6(2), 94–97.

Zaidi, E., Bachur, R., & Harper, M. (1999). Non-typhi Salmonella bacteremia in children. *Pediatr Infect Dis J* 18(12), 1073–1077.

Zaidi, M.B., Calva, J.J., Estrada-Garcia, T.E., Leon, V., Vasquez, G., Figueroa, G., Lopez, E., Contreras, J., Abbott, J., Zhao, S., McDermott, P., & Tolleson, L. (2008). Integrated food chain surveillance system for Salmonella spp in Mexico. *EID* 14(3), 429–435.

Where Does *Campylobacter* Come From? A Molecular Odyssey

Alison J. Cody, Frances M. Colles, Samuel K. Sheppard, and Martin C.J. Maiden

1 Introduction

Campylobacter is the most common bacterial cause of gastroenteritis, worldwide. Since the first description of the disease in the 1970s (Skirrow, 1977), the incidence of human campylobacteriosis in the UK, measured in terms of laboratory reports, has risen steadily, peaking at 57,674 reports in the year 2000; with 46,603 reports in 2006 (http://www.hpa.org.uk). Although generally self limiting, this disease has an important economic impact (Skirrow and Blaser, 1992). More serious complications, such as motor neurone paralysis, arise in 1–2 cases per 100,000 people in the UK and USA (Nachamkin et al., 1998). The disease also has an appreciable, yet less defined, impact in developing countries. Approximately 90% of human infection is caused by *C. jejuni*, with *C. coli* accounting for much of the rest (Gillespie et al., 2002).

    C. jejuni, and *C. coli* are Gram-negative curved rods with polar flagella and are highly motile (Ketley, 1997). In common with other members of the genus *Campylobacter* they are fastidious bacteria that are best isolated in a microaerophilic atmosphere using specific complex media, since they lack many of the genes needed to degrade carbohydrates or amino acids. Unlike most other *Campylobacter* species, they show optimum growth at 42°C, perhaps as a consequence of their association with avian species. It is thought that where campylobacters become environmentally stressed they enter a viable non-culturable state, in which case pre-incubation in an enrichment broth may help to recover the organism (Humphrey, 1989; Ketley, 1997).

    Campylobacter are ubiquitous, being commensal members of the gastrointestinal microbiota of poultry and other farm, animals, as well as many wild species. Such animal infection is rarely symptomatic, but provides sources of contamination, both directly and via the consumption of the afflicted animal. Run-off from farmland

M.C.J. Maiden (✉)

Department of Zoology, University of Oxford, South Parks Road, Oxford, OX1 2PS, UK

e-mail: martin.maiden@zoo.ox.ac.uk

A. Finn et al. (eds.), *Hot Topics in Infection and Immunity in Children VI*, Advances in Experimental Medicine and Biology 659, DOI 10.1007/978-1-4419-0981-7_4,

© Springer Science+Business Media, LLC 2010

leads to the contamination of ground-water sources, acting as a source of infection for animals, birds and humans, if consumed (Hopkins et al., 1984; Peabody et al., 1997). However, human campylobacteriosis is most commonly associated with the consumption of chicken or chicken products (Adak et al., 2005). Ingestion of *Campylobacter* by humans may cause disease resulting from the invasion of the intestinal epithelial layer, leading to localized inflammation and diarrhea (Young et al., 2007).

The multiple potential reservoirs for human infection, together with the high levels of genetic diversity of these bacteria, have combined to complicate understanding of the relative contributions of different infection sources to the human disease burden – an essential prerequisite to effective disease control. Here, we illustrate how molecular epidemiological techniques, especially those based on nucleotide sequence determination, have contributed to unravelling the epidemiology of these important human pathogens.

2 *Campylobacter* Typing and Population Structure

Multi-locus sequence typing (MLST) is a method of unambiguously indexing genetic variation among bacterial isolates (Maiden, 2006). For every isolate investigated, nucleotide sequence data are obtained for an internal fragment of each of seven housekeeping loci. Genetic variation in housekeeping genes is indexed as the variation is present in all isolates and is under stabilising selection for the conservation of metabolic function. This allows for the monitoring of long-term evolutionary events (Dingle et al., 2001). Nucleotide sequence lengths of $\sim$500 bp are usually employed, since accurate data can be readily obtained by the use of a single primer extension for each DNA strand. For each locus, every unique gene fragment sequence (or MLST allele) is assigned a unique but arbitrary number, regardless of whether allele differences have occurred as a result of a single or multiple base changes – an important criterion when analyzing highly recombinogenic bacteria (Maiden, 2006). Consequently, each isolate investigated has an "allelic profile" or "sequence type" (ST), consisting of seven integers which, in the case of the *Campylobacter* scheme, represents 3,309 bp of unique nucleotide sequence from multiple loci around the genome. Genetic relationships between isolates can be determined from these data since closely related isolates have identical STs or allelic profiles differing at few loci, while the profiles of unrelated isolates are different.

Examination of *C. jejuni* and *C. coli* collections by MLST has confirmed these organisms to be genetically diverse, with a semi-clonal population structure (Dingle et al., 2001, 2005). Such bacterial populations contain clusters of related isolates but, as in other highly recombinogenic bacteria, phylogenetic relationships among clusters cannot be determined, due to the reassortment of alleles by frequent recombination (Holmes et al., 1999). These clusters are referred to as clonal complexes and in the case of *Campylobacter* are pragmatically defined as those isolates sharing four or more alleles with a central genotype, after which the complex is named

(Dingle et al., 2001). As observed in other bacteria (Maiden, 2006), the central genotypes of *Campylobacter* clonal complexes have a higher prevalence than other STs in population samples and are stable during global spread over time (Dingle et al., 2002). An example of this is the ST-45 clonal complex, in which the central genotype is the most abundant; with the majority of other STs observed much less frequently, in many cases only once. Clonal complexes have become the main unit of analysis of *Campylobacter* genotypes for epidemiological investigations (Dingle et al., 2002).

3 Epidemiology of Human Infection

Prior to the development of MLST, immunological typing methods, predominantly targeted to lipooligosaccharide (LOS) variants and capsular antigens, were used for the characterization of *Campylobacter* isolates (Penner et al., 1983) but these methods failed to advance understanding of *Campylobacter* epidemiology. The reasons for this failure became clear on comparison of serotyping and MLST data. Firstly, particular serotypes can be associated with more than one ST or clonal complex; furthermore, a given genotype can contain isolates expressing various serotypes (Dingle et al., 2002; Wareing et al., 2003). This lack of association of serotype with genotype is due both to recombinational reassortment of antigen genes among genotypes and to phase variation, so that the same isolate may express very different serotypes at different times (Parkhill et al., 2000).

A major advantage of MLST is that the results are highly reproducible and the technique is portable, enabling data collected in different laboratories to be readily compared. MLST data available via the Internet at the PubMLST database website (http://pubmlst.org/Campylobacter) show that while *Campylobacter* genotyopes have a global distribution, their prevalence among human disease is not uniform worldwide (Dingle et al., 2008). The distribution of clonal complexes among disease isolates from two regions of the UK might be similar, for example (Dingle et al., 2002; Sopwith et al., 2006), but might differ from those seen among disease isolates collected in Australia (Mickan et al., 2007). Intriguingly, the disease isolates from the UK and Australia are more similar to each other than they are to isolates obtained from cases in the Dutch West Indies (Duim et al., 2003) (Fig. 1). These findings probably reflect different dietary infection sources in different countries. Of particular note is the similarity of the genotypes recovered from poultry meat and human disease, which is consistent with contaminated poultry being a major source of human infection (Dingle et al., 2002; Colles et al., 2008).

Outbreaks of human *Campylobacter* infection are identified infrequently due to the relatively long incubation period, lack of accurate epidemiological information, and the high incidence and wide distribution of human-disease related central genotypes. The application of a ten-locus typing scheme that combines MLST data and nucleotide sequences of *flaA*, *flaB* and *porA* antigen genes, enhances discrimination among isolates. In a study of 620 isolates collected over a one-year period from

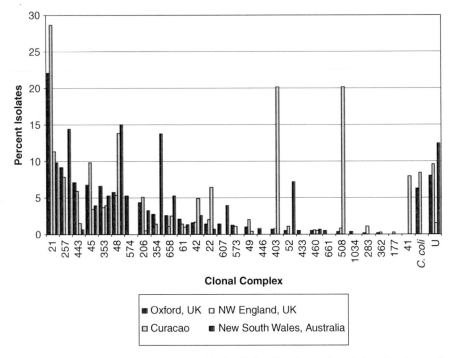

Fig. 1 Human clinical isolates in geographically distinct locations. The relative abundance of the clonal complexes detected in Oxfordshire, UK during a one-year study (Dingle et al., 2008). Comparison with the clonal complexes detected in three other studies of human *Campylobacter* infections in NW England, UK (Sopwith et al., 2006) Curaçao, (Duim et al., 2003) and New South Wales, Australia (Mickan et al., 2007)

Oxfordshire, UK the discriminatory index (DI) was increased from 0.976 achieved by MLST alone to 0.992 by achieved with ten-locus typing. This greatly enhanced the ability to detect outbreaks, although the contribution of such outbreaks to overall disease burden does, indeed, appear to be low (Dingle et al., 2008).

In addition to being useful in the study of human disease epidemiology, clonal complexes have provided information about the association of particular *Campylobacter* genotypes with particular animal hosts. In an analysis of 814 *Campylobacter* isolates from diverse sources, including farm and wild animals, as well as human disease, the ST-21 complex was isolated from all the sources; ST-45 and ST-257 complex predominantly from human disease or poultry; ST-61 and ST-48 complexes mainly from human disease, cattle and sheep, whereas ST-177 and ST-179 complexes were exclusively isolated from beach sand, presumably ultimately coming from wild birds. These findings suggest that the frequency distribution of some clonal complexes is in part associated with the environment from which isolates are obtained and that their persistence in particular hosts may be a result of niche adaptation (Dingle et al., 2002).

4 Host Association Studies of Chickens, Geese and Starlings

Understanding patterns of host association of different *Campylobacter* genotypes is important in elucidating the ecology of this food-borne zoonotic infection. If particular genotypes are associated with particular host species, it becomes possible to use genetic data to attribute cases of human infection to particular host sources. Initial investigations with MLST suggested that *Campylobacter* genotypes associated with human disease are found in the farm environment (Colles et al., 2003; French et al., 2005), indicating as obvious transmission routes to humans contaminated food that has been inadequately prepared.

Analysis of *Campylobacter* isolates obtained from 975 individual chickens representing 64 free-range broiler flocks, 331 wild geese and 964 wild starlings at the University farm in Oxfordshire between 2002 and 2005 enabled host association and transmission in a single farm to be investigated (Colles et al., 2008). There were some significant differences in the carriage rate and species distribution of the organism between the broiler chickens and wild birds. The average shedding rate was much higher amongst the broiler chickens (90.4%) compared to the wild geese (50.5%) and wild starlings (30.4%). In addition, *C. coli* was isolated much more commonly from the broiler chickens (50.9% of isolates) compared to the geese (0.6% of isolates) and starlings (1.7% of isolates). The results were consistent with those from other studies suggesting that *C. coli* is a later coloniser of chickens than *C. jejuni* and is thus commonly isolated from free-range fowl and poultry organically reared over a longer period of time (El-Shibiny et al., 2005).

Further comparison of the *Campylobacter* genotypes recovered revealed strong host association, with little overlap between the different host sources. Only five clonal complexes and four STs overlapped amongst the chickens and wild birds, and only two *Campylobacter* isolates with identical ST and *flaA* SVR type were isolated from the chickens and starlings, although the time period was 6 months apart. Both Fisher's statistic, used to compare the gene flow between *Campylobacter* populations, and analysis using the software package *Structure* (Falush et al., 2003) to predict the host source from an allelic profile employing probabilistic methods, supported a strong association of particular *Campylobacter* genotypes with certain hosts (Fig. 2). This finding was consistent with mapping genealogical data to host source,

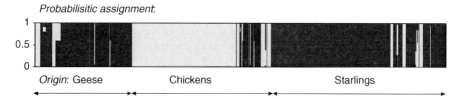

Fig. 2 Probabilistic assignment of the *Campylobacter jejuni* allelic profiles isolated from geese (*blue*), chickens (*yellow*) and starlings (*red*), using the software package *Structure*. Each vertical bar represents an allelic profile and gives the estimated probability that it comes from each of the sources

which shows an overrepresentation of the same genotypes in farm – specifically poultry – sources and human disease (Fig. 3). This association of genotypes with hosts and human disease supports the use of genetic methods to determine major sources of human infection.

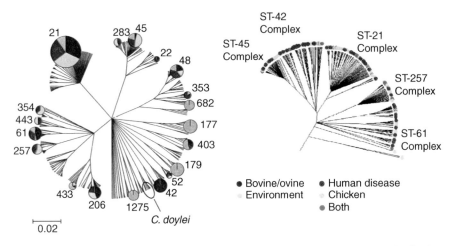

Fig. 3 Comparison of human *Campylobacter jejuni* isolates with those from poultry, ovine/bovine and environmental sources. The phylogenetic trees were constructed using ClonalFrame software and the ST-based clonal complexes are indicated. The shaded area of the pie charts is proportional to the frequency that each clonal complex occurs in comparison of scaled data sets ($n = 765$)

Molecular epidemiology can also be used to track infection on-farm. The free-range poultry flocks examined in the Oxfordshire study were on a rolling production system using 16 different plots at two different farm sites. Plots were fallow for 7 weeks between successive flocks and a total of four flocks was raised per plot over the study period. The number of occasions in which an identical *Campylobacter* genotype was identified on the same plot was unpredictable and no different to that which might expected by chance, despite the inevitable contamination that would result from the previous flocks. Similarly, the extent of ranging behaviour showed no correlation with the shedding rate of *Campylobacter* or the diversity of geno-types. Higher rates of genotypic diversity were, however, linked with lower growth rates and better hock health. Genotypes were clustered over time, independently of farm site and plot, suggesting that the flocks were contaminated from a common source over a period of time. Taken together these data suggest that wild birds are not a common source of contamination for the free-range poultry but imply that *Campylobacter* genotypes may be circulating within the poultry industry (Colles et al., 2008).

5 Genetics of Host Association and Speciation

A further application of MLST data has been the investigation of the evolutionary processes underlying host adaptation. A particular conundrum is how distinct genotypes can emerge in the face of the high rates of recombination in *Campylobacter* populations. Specifically, the persistence of different host-adapted genotypes, and indeed the two species, seem to be inconsistent with the apparently high rates of genetic exchange (Fraser et al., 2007). The existence of distinct groups suggests the presence of effective barriers to gene flow (Lawrence, 2002). This can be (i) mechanistic – imposed by homology dependence of recombination (Fraser et al., 2007) or other factors promoting DNA specificity (Eggleston and West, 1997) such as restriction/modification systems; (ii) ecological – a consequence of physical separation of bacterial populations in distinct niches; (iii) adaptive – implying selection against hybrid genotypes (Zhu et al., 2001). For many bacteria, observed levels of genetic exchange should prevent divergence of subpopulations (Gupta and Maiden, 2001; Cohan, 2002; Hanage et al., 2006).

In *Campylobacter* there is evidence of the mechanisms of adaptation and it has been suggested that *C. jejuni* and *C. coli* are despeciating as a result of expansion into an agricultural niche (Sheppard et al., 2008). As livestock farming became an increasingly important part of human food production, natural barriers to bacterial recombination became less rigid. Animals with undomesticated predecessors, from whom they were separated in the wild, were brought together at high stocking densities with rapid generation times. The convergence of *Campylobacter* species in this new niche demonstrates that bacterial evolution can occur by mechanisms analogous to those in sexual eukaryotic populations, such as Darwin's Finches, where ecological factors can generate and maintain incipient species associated with distinct niches (Grant and Grant, 1992; Mallet, 2007).

6 Conclusion – Impact of Molecular Epidemiology on Control of Human Infection

Molecular epidemiological studies have contributed to our understanding of the biology of *Campylobacter* infection, and indeed help to elucidate where *Campylobacter* comes from, in a number of important ways. MLST and related typing techniques have provided a robust typing tool that has enabled data from multiple laboratories to be compared. The clonal complex has emerged as a practical, yet biologically based, unit of epidemiological analysis. Enhancement of these data with additional loci encoding more variable genes has provided robust tools for contact tracing and outbreak investigation. MLST has also enabled the association of particular genotypes with particular host species, further confirming the important role of poultry as a source of human disease and indicating that the animals, themselves, have *Campylobacter* populations distinct from those found among wild birds such as geese and starlings. Finally the data have helped us to understand the evolutionary processes whereby distinct genotypes arise and are maintained.

In practical terms, much emphasis has been placed on reducing the number of *Campylobacter* organisms reaching the consumer, since is directly proportional to the risk of infection (Newell and Fearnley, 2003; Lindqvist and Lindblad, 2008). In order to achieve this reduction, research has largely concentrated on three main areas; on-farm biosecurity, hygiene during the slaughter process and hygiene during food preparation. Although some progress has been made, further improvement is needed to meet the target set by the UK Food Standards Agency in 2005 to achieve a 50% reduction in *Campylobacter*-infected UK-produced chickens by 2010. The results of the molecular epidemiological studies suggest that two additional areas could be targeted in the future: improved animal welfare and the impact of intensive farming (Fig. 4) (Humphrey, 2006; Colles et al., 2008; Sheppard et al., 2008). This illustrates how molecular epidemiological and population genetic studies are important in shedding light on the best means to reduce the incidence of this major pathogen.

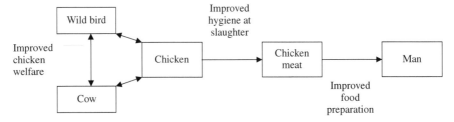

Fig. 4 Schematic representation of intervention points for reducing rates of human infection, suggested by molecular epidemiology investigations

References

Adak G.K., Meakins S.M., Yip H., Lopman B.A., & O'Brien S.J. (2005). Disease risks from foods, England and Wales, 1996–2000. *Emerg Infect Dis*, (11), 365–372.

Cohan F.M. (2002). What are bacterial species? *Annu Rev Microbiol*, (56), 457–487.

Colles F.M., Jones K., Harding R.M., & Maiden M.C. (2003). Genetic diversity of *Campylobacter jejuni* isolates from farm animals and the farm environment. *Appl Environ Microbiol*, (69), 7409–7413.

Colles F.M., Jones T.A., McCarthy N.D., Sheppard S.K., Cody A.J., Dingle K.E., Dawkins M.S., & Maiden M.C. (2008). *Campylobacter* infection of broiler chickens in a free-range environment. *Environ Microbiol*, (10), 2042–2050.

Dingle K.E., Colles F.M., Falush D., & Maiden M.C. (2005). Sequence typing and comparison of population biology of *Campylobacter coli* and *Campylobacter jejuni*. *J Clin Microbiol*, (43), 340–347.

Dingle K.E., McCarthy N.D., Cody A.J., Peto T.E., & Maiden M.C. (2008). Extended sequence typing of *Campylobacter* spp. *Emerg Infect Dis*, (14), 1620–1622.

Dingle K.E., Colles F.M., Ure R., Wagenaar J., Duim B., Bolton F.J., Fox A.J., Wareing D.R.A., & Maiden M.C.J. (2002). Molecular characterization of *Campylobacter jejuni* clones: a rational basis for epidemiological investigations. *Emerg Infect Dis*, (8), 949–955.

Dingle K.E., Colles F.M., Wareing D.R.A., Ure R., Fox A.J., Bolton F.J., Bootsma H.J., Willems R.J.L., Urwin R., & Maiden M.C.J. (2001). Multilocus sequence typing system for *Campylobacter jejuni*. *J Clin Microbiol*, (39), 14–23.

Duim B., Godschalk P.C., van den Braak N. et al. (2003). Molecular evidence for dissemination of unique *Campylobacter jejuni* clones in Curacao, Netherlands Antilles. *J Clin Microbiol*, (41), 5593–5597.

Eggleston A.K. & West S.C. (1997). Recombination initiation: easy as A, B, C, D... chi? *Curr Biol* (7), R745–R749.

El-Shibiny A., Connerton P.L., & Connerton I.F. (2005). Enumeration and diversity of campylobacters and bacteriophages isolated during the rearing cycles of free-range and organic chickens. *Appl Environ Microbiol*, (71), 1259–1266.

Falush D., Stephens M., & Pritchard J.K. (2003). Inference of population structure using multilocus genotype data: linked loci and correlated allele frequencies. *Genetics*, (164), 1567–1587.

Fraser C., Hanage W.P., & Spratt B.G. (2007). Recombination and the nature of bacterial speciation. *Science*, (315), 476–480.

French N., Barrigas M., Brown P., Ribiero P., Williams N., Leatherbarrow H., Birtles R., Bolton E., Fearnhead P., & Fox A. (2005). Spatial epidemiology and natural population structure of *Campylobacter jejuni* colonizing a farmland ecosystem. *Environ Microbiol*, (7), 1116–1126.

Gillespie I.A., O'Brien S.J., Frost J.A., Adak G.K., Horby P., Swan A.V., Painter M.J., Neal K.R., & Collaborators C.S.S.S. (2002). A case-case comparison of *Campylobacter coli* and *Campylobacter jejuni* infection: a tool for generating hypotheses. *Emerg Infect Dis*, (8), 937–942.

Grant P.R. & Grant B.R. (1992). Hybridization of bird species. *Science*, (256), 193–197.

Gupta S. & Maiden M.C.J. (2001). Exploring the evolution of diversity in pathogen populations. *Trends Microbiol*, (9), 181–192.

Hanage W.P., Spratt B.G., Turner K.M., & Fraser C. (2006). Modelling bacterial speciation. *Phil Trans Roy Soc Lond B Biol Sci* (361), 2039–2044.

Holmes E.C., Urwin R., & Maiden M.C.J. (1999). The influence of recombination on the population structure and evolution of the human pathogen *Neisseria meningitidis*. *Mol Biol Evol*, (16), 741–749.

Hopkins R.S., Olmsted R., & Istre G.R. (1984). Endemic *Campylobacter jejuni* infection in Colorado: identified risk factors. *Am J Pub Health*, (74), 249–250.

Humphrey T. (2006). Are happy chickens safer chickens? Poultry welfare and disease susceptibility. *Br Poult Sci*, (47), 379–391.

Humphrey T.J. (1989). An appraisal of the efficacy of pre-enrichment for the isolation of *Campylobacter jejuni* from water and food. *J Appl Bacteriol*, (66), 119–126.

Ketley J.M. (1997). Pathogenesis of enteric infection by *Campylobacter*. *Microbiology*, (143), 5–21.

Lawrence J.G. (2002). Gene transfer in bacteria: speciation without species? *Theor Popul Biol*, (61), 449–460.

Lindqvist R. & Lindblad M. (2008). Quantitative risk assessment of thermophilic *Campylobacter* spp. and cross-contamination during handling of raw broiler chickens evaluating strategies at the producer level to reduce human campylobacteriosis in Sweden. *Int J Food Microbiol* (121), 41–52.

Maiden M.C. (2006). Multilocus sequence typing of bacteria. *Annu Rev Microbiol*, (60), 561–588.

Mallet J. (2007). Hybrid speciation. *Nature*, (446), 279–283.

Mickan L., Doyle R., Valcanis M., Dingle K.E., Unicomb L., & Lanser J. (2007). Multilocus sequence typing of *Campylobacter jejuni* isolates from New South Wales, Australia. *J Appl Microbiol*, (102), 144–152.

Nachamkin I., Allos B.M., & Ho T. (1998). *Campylobacter* species and Guillain-Barré syndrome. *Clin Microbiol Rev*, (11), 555–567.

Newell D.G. & Fearnley C. (2003). Sources of *Campylobacter* colonization in broiler chickens. *Appl Environ Microbiol*, (69), 4343–4351.

Parkhill J., Wren B.W., Mungall K. et al. (2000). The genome sequence of the food-borne pathogen *Campylobacter jejuni* reveals hypervariable sequences. *Nature*, (403), 665–668.

Peabody R., Ryan M.J., & Wall P.G. (1997). Outbreaks of *Campylobacter* infection: rare events for a common pathogen. *CDR*, (7), R33–R37.

Penner J.L., Hennessy J.N., & Congi R.V. (1983). Serotyping of *Campylobacter jejuni* and *Campylobacter coli* on the basis of thermostable antigens. *Eur J Clin Microbiol*, (2), 378–383.

Sheppard S.K., McCarthy N.D., Falush D., & Maiden M.C. (2008). Convergence of *Campylobacter* species: implications for bacterial evolution. *Science*, (320), 237–239.

Skirrow M.B. (1977). Campylobacter enteritis: a "new" disease. *BMJ*, (2), 9–11.

Skirrow M.B. & Blaser M.J. (1992). *Clinical and Epidemiological Considerations*. Tompkins LS (Ed.), pp. 3–8. Washington, DC: American Society for Microbiology.

Sopwith W., Birtles A., Matthews M., Fox A., Gee S., Painter M., Regan M., Syed Q., & Bolton E. (2006). *Campylobacter jejuni* multilocus sequence types in humans, northwest England, 2003–2004. *Emerg Infect Dis*, (12), 1500–1507.

Wareing D.R., Ure R., Colles F.M., Bolton F.J., Fox A.J., Maiden M.C., & Dingle K.E. (2003). Reference isolates for the clonal complexes of *Campylobacter jejuni*. *Lett Appl Microbiol*, (36), 106–110.

Young K.T., Davis L.M., & Dirita V.J. (2007). *Campylobacter jejuni:* molecular biology and pathogenesis. *Nat Rev Microbiol*, (5), 665–679.

Zhu P., van der Ende A., Falush D. et al. (2001). Fit genotypes and escape variants of subgroup III *Neisseria meningitidis* during three pandemics of epidemic meningitis. *Proc Natl Acad Sci USA*, (98), 5234–5239.

Why Are Some Babies Still Being Infected with HIV in the UK?

Aubrey Cunnington, Sanjay Patel, and Hermione Lyall

1 Introduction

The risk of transmission of HIV from a mother to her baby can be reduced from 35%, in the absence of interventions, to 0.1% with optimal management and exclusion of breastfeeding. However, the reality is that transmission rates in the UK and Ireland are 10 fold higher at approximately 1%. The underlying reasons are complex, but mainly they reflect a failure to detect all women with HIV in pregnancy and system failures that result in missed opportunities for risk reduction. Here we review the mechanisms and risk factors for mother to child transmission (MTCT) of HIV, the current guidelines for prevention of MTCT, and the results and recommendations from a recent study of cases of MTCT in the UK. Finally, we use an instructive clinical scenario to illustrate the need for all healthcare workers to be aware of the guidelines for prevention of MTCT of HIV when dealing with pregnant women and their babies.

2 Risk Factors for Transmission

In the absence of any interventions, up to 35% of breastfed infants born to HIV-1 infected mothers will acquire HIV infection (Peckham and Gibb 1995; Newell, 1998; Lehman and Farquhar 2007). HIV can be transmitted by the passage of cell free or cell-associated virus from mother to child (John et al., 2001; Tuomala et al., 2003; Koulinska et al., 2006); it is present in maternal blood, vaginal secretions, and breast milk (Dickover et al., 1996; Nielsen et al., 1996; Tuomala et al., 2003; Koulinska et al., 2006), and it may pass into the fetus by direct transmission across the placenta or through contact with epithelial surfaces (Bryson et al., 1992; Mandelbrot et al., 1996; Nielsen et al., 1996; Kwiek et al., 2006). The infant may also be infected by ingestion of breast milk.

A. Cunnington (✉)

Immunology Unit, London School of Hygiene and Tropical Medicine, London, UK

e-mail: acunning@doctors.org.uk

A. Finn et al. (eds.), *Hot Topics in Infection and Immunity in Children VI*, Advances in Experimental Medicine and Biology 659, DOI 10.1007/978-1-4419-0981-7_5,

© Springer Science+Business Media, LLC 2010

The timing of vertical infection is usually considered as either in utero, intrapartum or postpartum (due to breastfeeding) (Bryson et al., 1992; Lehman and Farquhar 2007). It is difficult to be precise about the timing or mechanisms of vertical infection around delivery because many different exposures occur within a short time period. Differentiating between late in utero, intrapartum, and early breastfeeding transmission is impossible in practice. However, observational studies, mathematical modeling, and data derived from intervention studies suggest that in utero infection is uncommon but increases toward term; intrapartum transmission accounts for around two-thirds of infections; and breastfeeding accounts for the remainder (Peckham and Gibb 1995; Newell 1998; Lehman and Farquhar 2007). In the absence of interventions, 5–10% of infants would be infected in utero, 10–20% intrapartum, and 10–20% by breastfeeding. Conventionally, infants who test HIV DNA PCR positive in the first 48 h of life are considered to have been infected in utero, while those who are infected at birth are usually still HIV DNA PCR negative in peripheral blood lymphocytes at this time (Bryson et al., 1992).

The risk of transmission of HIV to the fetus or infant is influenced by the intensity and duration of exposure to HIV and how readily the virus can cross the placental or epithelial barriers. Advanced maternal disease with high viral load (VL) in maternal plasma is a consistent risk factor for transmission of HIV in utero, intrapartum, and through breastfeeding (Dickover et al., 1996; John et al., 2001; Rousseau et al., 2003; Coutsoudis et al., 2004). Similarly, viral loads in vaginal fluid and in breast milk are correlated with the risk of transmission, and the presence of HIV in amniotic fluid is likely to increase the risk of in utero infection (Mandelbrot et al., 1996; John et al., 2001; Rousseau et al., 2003). Other infections affecting the mother may increase local concentrations of cell free or cell associated HIV. Thus chorioamnionitis and sexually transmitted infections during pregnancy, active genital herpes, other ulcerative sexually transmitted infections at the time of delivery, and mastitis during breast feeding have all been reported to increase transmission of HIV(Burns et al., 1994; Landesman et al., 1996; John et al., 2001).

In utero infection may occur due to HIV crossing the placenta and entering the fetal circulation, or through amniotic fluid bringing virus into contact with fetal mucosal surfaces. Disruption of the integrity of the placenta or amniotic membranes probably increases the risk of transmission (Mandelbrot et al., 1996). Smoking and illicit drug use can compromise the placental integrity, as can amniocentesis or physical damage to the placenta (Burns et al., 1994; Landesman et al., 1996; Mandelbrot et al., 1996). Other acute infections in pregnancy (e.g. malaria or syphilis), which compromise the integrity of the placenta, may also increase the risk of in utero transmission (Mwapasa et al., 2006; Brahmbhatt et al., 2008).

The risk of intrapartum infection is clearly influenced by the duration of rupture of membranes, with the relative risk of transmission increasing by approximately 2% per hour during the first 24 h (Landesman et al., 1996; International Perinatal HIV Group, 2001). Premature delivery is an independent risk factor for transmission (Landesman et al., 1996; Mandelbrot et al., 1996; Kuhn et al., 1997). Transplacental materno-fetal microtransfusions, which can be quantified by measuring placental alkaline phosphatase or maternal DNA in umbilical cord blood, show a dose-response relationship with risk of transmission (Kwiek et al., 2006, 2008). Maternal haemorrhage during labour and damage to fetal skin also increase the chance of transmission (Mandelbrot et al., 1996; Maiques et al., 1999).

The risk of transmission associated with breast feeding is related to the duration of breast feeding (Coutsoudis et al., 2004; Coovadia et al., 2007). Almost two-thirds of transmission by this route occurs in the first 6 weeks. Thereafter the risk per month of breastfeeding is lower but remains fairly constant. Exclusive breast feeding is thought to be safer than mixed breast and bottle feeding, perhaps because gut mucosal integrity is impaired by introduction of other feeds, which may facilitate entry of HIV (Kourtis et al., 2003).

3 British HIV Association Guidelines – 2008

The remainder of this review concentrates on the prevention of MTCT of HIV in a resource rich setting, focusing primarily on the UK. Further details can be found in the British HIV Association guidelines (de Ruiter et al., 2008). The standard of care expected for a pregnant HIV infected woman and her newborn child has greatly exceeded the evidence available from randomised controlled trials of interventions but is based on simple general principles and supported by the impressive declines observed in a number of countries in the rate of mother to child transmission.

The first and most important step in this process is the aim to diagnose every pregnant woman who has HIV. Since 2000, routine HIV testing during pregnancy has been offered in the UK. Therefore all women, at the start of their antenatal care, should be recommended antenatal testing for HIV alongside testing for other infections (syphilis, hepatitis, rubella). Women who decline and those at ongoing risk of infection should be offered testing again later in pregnancy. Women presenting without previous antenatal care late in pregnancy or at delivery should be offered a rapid test so that treatment can be initiated immediately.

The health of the mother must be considered and treatment should be started if she requires it either because of symptomatic HIV or a low CD4 count. If the mother requires treatment, it should be highly active antiretroviral therapy (HAART), ideally begun after the first trimester. If she does not require treatment for her own health, then treatment options to reduce transmission to the infant include either of the following: zidovudine monotherapy with planned pre-labour caesarean section for women with CD4 counts >200 cells/mm^3 and HIV VL <5,000–10,000 copies/ml; or short-term combination antiretroviral therapy (START) for any CD4 count or VL. With START, and adherence to treatment, the VL should be <50 copies/ml before delivery, and, in this instance, women can choose whether to go ahead with a vaginal birth.

Choice of antiretrovirals in pregnancy should take account of potential toxicities to both mother and fetus and whether transplacental passage of the drugs is desirable. Potential teratogenic effects of HAART exposure are prospectively monitored through voluntary reporting to the international antiretroviral pregnancy registry at the following address: www.apregistry.com. To date, the overall rate of congenital abnormalities for infants born to mothers who were exposed to HAART in the first trimester is equivalent to the background population rate of 3% (Antiretroviral Pregnancy Registry Steering Committee, 2008). A slight increase in a number of different abnormalities has been noted following didanosine exposure (4–5%), but this drug is rarely used in pregnant women now. Due to the teratogenic effects of efavirenz on the central nervous system (noted in 3 of 20 efavirenz-treated cynomolgus monkeys compared to 0 of 20 untreated controls), this drug has been assigned an FDA category D (positive evidence of risk in pregnancy) and contra-indicated for pregnant women and women who are of childbearing age. However, to date, of 407 pregnancies reported to the registry after first trimester exposures to efavirenz, there have been only 13 (3.2%) congenital abnormalities, including only one case each of myeolmenigocele and anopthalmia. All women who conceive on any combination of HAART should be recommended to have a fetal ultrasound scan at 20 weeks to detect anomalies. Maternal HAART has been demonstrated to increase the risk of pre-term delivery in a number of European studies (Thorne et al., 2004; Townsend et al., 2007).

Antiretrovirals differ in their passage across the placenta. Nevirapine has excellent transfer across the placenta and a long half life such that even a single dose given to the mother at least 2 h before delivery provides an effective loading dose to the infant, which remains at a therapeutic level for up to 7 days after delivery (Musoke et al., 1999). The nucleosides zidovudine, lamivudine, and tenofovir also cross the placenta efficiently; however, most of the protease inhibitors do not cross the placenta to any significant extent (Pacifici, 2006; Hirt et al., 2009).

Maternal plasma VL at 36 weeks gestation is an important guide for the management of the delivery and the neonate. If the VL is detectable, then delivery by pre-labour Caesarean section is advised, as is the case for mothers receiving zidovudine monotherapy. Pre-labour Caesarean section is also advised for women with adherence difficulties or resistant virus where complete suppression of virus may not be possible.

Management of complications of pregnancy must balance risk of transmission with risks of interventions to mother and fetus. This is particularly important with the management of preterm rupture of membranes where expedited delivery may carry a significant risk of morbidity and mortality associated with prematurity, but delay increases the risk of HIV transmission, especially if the mother is not on HAART and has a detectable VL.

Finally, management of the newborn is determined by whether maternal VL is fully suppressed (<50 copies/ml) or not and by events at delivery. If delivery does not occur as planned and there is increased risk of transmission such as spontaneous delivery before the date of planned pre-labour Caesarean section, then the infant should receive post-exposure prophylaxis (PEP) with three antiretrovirals rather than the planned single drug treatment (which is only sufficient when maternal VL is fully suppressed). Figure 1a, b summarise the current UK guidelines, although reference regarding patient management should be made to the comprehensive British HIV Association guidelines (de Ruiter et al., 2008).

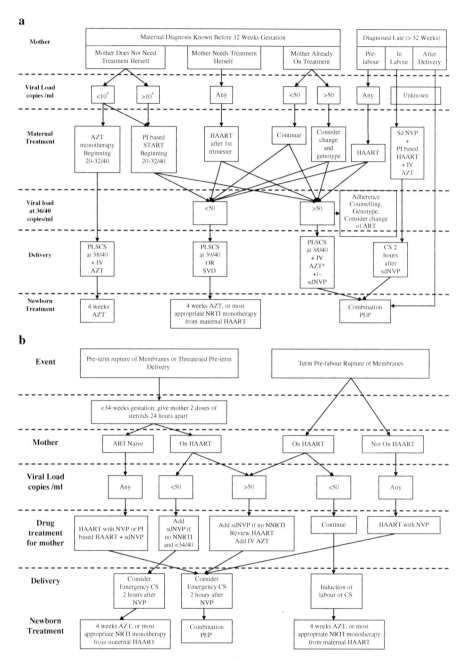

Fig. 1 Flowcharts to summarise the 2008 BHIVA guidelines for the management of HIV in pregnancy. ART, antiretroviral therapy; AZT, zidovudine; CS, caesarean section; HAART, highly active antiretroviral therapy; IV, intravenous; NNRTI, non-nucleoside reverse transcriptase inhibitor; NRTI, nucleoside reverse transcriptase inhibitor; PEP, post exposure prohylaxis; PI, protease inhibitor; PLSCS, pre-labour caesarean section; START, short term antiretroviral therapy; sd NVP, single dose nevirapine; SVD, spontaneous vaginal delivery, *if no evidence of resistance to zidovudine

4 Transmission in the Era of HAART

Strategies to reduce mother to child transmission have evolved, as risk factors for transmission have been identified and new antiretroviral drugs have become available (Foster and Lyall 2006). These developments resulted in the current low rates of perinatal transmission (Fig. 2) (Townsend et al., 2008a, b).

Fig. 2 Use of antiretroviral therapy in pregnancy by year of delivery in the UK and Ireland. Printed with permission of Dr Pat Tookey

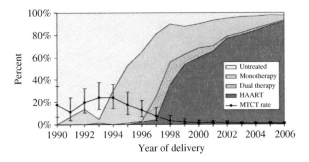

Data from the UK and French perinatal cohorts demonstrate that the longer the mother has been treated with HAART the lower the risk of transmission, allowing vertical transmission rates to be brought down as low as 0.1–0.4% in infants with maternal HIV-1 VL <50 copies/ml (Townsend et al., 2008a; Warszawski et al., 2008). Table 1 shows transmission rates for the UK cohort; however, the overall rate of vertical transmission in infected women delivering in developed countries remains around 1%.

Table 1 UK cohort data for mother to child transmission of HIV (Townsend et al., 2008a). MTCT, mother to child transmission; ART, antiretroviral therapy; HAART, highly active antiretroviral therapy; CS, caesarean section; ZDV mono, zidovudine monotherapy

	MTCT rate	95% CI	n infected	Total
Overall	1.2	(0.9–1.5)	61	5151
2000–2002	1.6	(1.0–2.4)	23	1456
2003–2006	1.0	(0.7–1.4)	38	3695
At least 14 days of ART	0.8	(0.6–1.1)	40	4864
ART and mode of delivery				
HAART + elective CS	0.7	(0.4–1.2)	17	2337
HAART + planned vaginal	0.7	(0.2–1.8)	4	565
HAART + emergency CS	1.7	(1.0–2.8)	15	877
ZDV mono + elective CS	~0.0	(0.0–0.8)	0	467
HAART	1.0	(0.7–1.3)	40	4120
HAART from conception	0.1	(0.0–0.6)	1	928
HAART + VL<50 copies/ml	0.1	(0.0–0.4)	3	2117

The reasons for ongoing infant HIV infection, despite high antenatal detection rates (>95% before delivery) and a high uptake of effective interventions (>95%), were specifically addressed in The Audit for Perinatal Transmission of HIV in

England, 2002–2005 (Anonymous, 2007). This study investigated cases of vertically infected children and identified potential risk factors for transmission, aiming to provide recommendations to reduce future transmission rates. The impact of failing to diagnose women prior to delivery was clearly demonstrated in this study. Fifty-four of the 87 children with HIV were born to women whose HIV infection was undiagnosed. Mortality in the first 2 years of life from an AIDS related condition was 17% in the undiagnosed group, with 60% of the survivors having an AIDS defining illness as compared to a 6% mortality rate in the diagnosed group. Although more than a third of these 54 mothers declined HIV testing during pregnancy, at least 20% of them are likely to have seroconverted during pregnancy, indicating a possible role for repeat HIV testing in the third trimester as well as a need for good sexual health advice for pregnant and lactating women.

How can mothers who initially decline HIV testing be encouraged to reconsider their decision? Patients with complex social needs should be discussed within a multidisciplinary forum, and repeat offers of the test should be made by experienced members of the team; social support should also be offered where required. Obviously, the team's principal aim is to identify infected mothers prior to the third trimester, enabling timely access to treatment to achieve viral suppression prior to delivery. If this identification has not been achieved, even making a diagnosis during labour can allow interventions, which can both significantly reduce the rate of MTCT and also identify babies who will need to be followed up (Nesheim et al., 2007). Point of care HIV tests should be available on all labour wards (Rahangdale and Cohan 2008). These should be used for untested or unbooked mothers prior to delivery as well as for at risk mothers who may have contracted HIV during pregnancy. Assuming that a mother who tested negative at booking is still negative during labour may be a costly mistake. Partner testing (Walmsley, 2003) and repeat testing of pregnant women in the third trimester also have the potential to reduce transmission rates, although cost-benefit analysis for both of these interventions is required before introducing them as a routine measure. It is important to ensure that HIV positive pregnant asylum seekers and their infants receive all appropriate interventions to prevent MTCT (British Medical Association, 2008).

The audit findings in the 33 infected babies born to mothers diagnosed with HIV prior to delivery are also concerning. In a number of cases, system failures such as delays in antenatal testing, laboratory delays in reporting positive HIV results, delays in initiating antiretroviral treatment, and inadequate diagnosis and management of concurrent sexually transmitted infections may have contributed to babies being vertically infected. The findings from the audit underline the importance of providing patients with results in a timely manner, appropriate management of suboptimal viral loads, and the establishment of well defined communication pathways between healthcare professionals. To achieve these essential standards, each region needs to accept its collective responsibility by having a named HIV lead clinician for both obstetrics and paediatrics responsible for putting birth plans into place and also for ensuring that the management of each pregnancy complies with current national guidelines (de Ruiter et al., 2008). Effectively implementing the current national guidelines is a priority to reduce vertical transmission to the lowest possible rate.

Finally, earlier use of HAART has been proposed to reduce in utero transmission (Warszawski et al., 2008). Although transmission seems to be significantly reduced if HAART is used for at least 12 weeks, the benefits of extending the length of antenatal HAART beyond this period needs to be carefully evaluated against the risks of toxicities such as mitochondrial dysfunction, neonatal malformations, and prematurity (Tuomala et al., 2002; Watts et al., 2004).

In summary, in the UK, despite the universal availability of HAART for pregnant women with HIV, the rate of vertical transmission remains relatively static at 1%. To improve this rate, we must implement a robust system to ensure correct and consistent adherence to the current British HIV Association (BHIVA) guidelines. Once this has been achieved, further changes may need to be considered to further reduce the rate of vertical transmission.

5 Clinical Scenario – Emergency Management of a New Diagnosis of Maternal HIV at Delivery

You are called by the labour ward coordinator because a mother who has had no prior antenatal care has presented in early labour at term. The mother arrived in the UK from Sub-Saharan Africa about 6 months ago. A HIV point-of-care test has just been done and the result is positive.

5.1 What Should be Done Before the Baby Is Born?

While a laboratory HIV test should confirm the result of the point-of-care test, a false positive result is less likely in a high-risk case such as this one. In this emergency situation, the team should assume that the woman has HIV, as there is insufficient time to wait for a confirmatory result. The mother should be informed of the result of her test as sensitively as possible and advised that interventions are necessary to reduce the risk of transmission to the baby. A short history should be taken to determine the following: if she had in fact been aware of her diagnosis previously, if she had ever received antiretroviral treatment and whether she may have any other infections which may pose a risk to her baby. Confirmatory HIV serology should be requested as well as hepatitis B, C, syphilis, and rubella serology. A baseline sample should be sent for HIV viral load, HIV resistance testing, and CD4 count.

The mother should be commenced urgently on protease inhibitor based HAART and receive a stat dose of nevirapine and intravenous zidovudine. The mother should be counseled that an emergency Caesarean section at least 2 h after she has taken nevirapine is the safest option for her baby (Fig. 1b).

5.2 Which Antiretrovirals Should be Started for the Baby?

Post-exposure prophylaxis (PEP) must be started for the baby as soon as possible after delivery. If the mother is treatment naïve, then the current recommendation is to start the baby on zidovudine, lamivudine, and nevirapine (de Ruiter et al., 2008). The neonatal doses for these antiretrovirals and their side-effects are listed in Table 2. Nevirapine is given at a lower dose for the first week, which is then increased in the second week to compensate for cytochrome p450 induction that increases its metabolism. Nevirapine is stopped after 2 weeks while the other two antiretrovirals continue for the full 4 weeks. This staggered stop is to cover the long half life of nevirapine, so avoiding a period of monotherapy and consequent risk of inducing resistance (Mackie et al., 2004).

If the mother is known or strongly suspected to have resistant virus, then the neonate needs to be treated with a regimen that will effectively treat this resistant virus. Urgent expert advice should be obtained, because for many drugs there is very little pharmacokinetic data for neonates, and new classes of drugs may be used with only pharmacokinetic data from adult studies as a guide.

Table 2 Antiretroviral drug doses for neonates and possible side effects (de Ruiter et al., 2008)

Drug	Dose	Side effects
Zidovudine (AZT/ZDV)	**Oral** *Term (>34 weeks)* 4 mg/kg bd 2 mg/kg qds *Premature (30–34 weeks)* 2 mg/kg bd for 2 weeks then 2 mg/kg tds for 2 weeks *Premature (<30 weeks)* 2 mg/kg bd for 4 weeks **Intravenous** *Term (>34 weeks)* 1.5 mg/kg qds *Premature (<34 weeks)* 1.5 mg/kg bd	Anaemia, neutropenia Concerns about mitochondrial toxicity
Lamivudine (3TC)	*Baby* 2 mg/kg bd	Anaemia, neutropenia Concerns about mitochondrial toxicity Asymptomatic transient hyperlactataemia
Nevirapine (NVP)	*Mother* 200 mg in labour *Baby* 2 mg/kg od for first week followed by 4 mg/kg od for second week, then stop (use 4 mg/kg od for 2 weeks if the mother has received >3 days NVP)	
Co-trimoxazole	<6 months 120 mg od Mon/Wed/Fri	Rash, bone marrow suppression (especially neutropenia)

5.3 What Should be Done if the Mother is in Premature Labour?

Again, the mother should receive a single dose of nevirapine at least 2 h prior to delivery to provide a loading dose for the fetus. Intravenous zidovudine should also be administered, and the mother should be started on HAART. There are no contraindications to giving antenatal steroids to promote lung maturation if the gestation is less than 34 weeks (de Ruiter et al., 2008). The decision about whether or not to deliver by emergency-Caesarean section or to allow the baby to be delivered vaginally depends on the obstetric considerations, on the gestation and size of the baby. In view of the high risk of infection, the baby should ideally be started on PEP; however, often the degree of prematurity prevents oral medication from being given, and zidovudine is the only antiretroviral that is available for sick or premature babies who are unable to tolerate oral medication (Capparelli et al., 2003).

5.4 How Should the Infant be Tested for HIV?

The gold standard for diagnosing HIV in infancy is HIV proviral DNA PCR (sent in an EDTA sample), as opposed to HIV serology which is used in adults. Neonates should be tested on the first day of life to distinguish between in utero and intrapartum transmission. In addition, it is advised to send an accompanying maternal sample at this time to confirm that the primers being used are able to amplify the maternal virus. If this is found not to be the case, then a different primer set must be used to avoid falsely reassuring negative results in the baby. Neonatal ART can delay the detection of HIV DNA and therefore a second HIV test is performed at 6 weeks of age, 2 weeks after neonatal PEP has been stopped (Prasitwattanaseree et al., 2004). If the third PCR at 12 weeks in a non-breastfed baby is also negative, then the parents can be reassured that the baby is very unlikely to be infected. Absence of HIV infection in the infant is confirmed with a negative HIV antibody result at 18 months of age.

In a high-risk infant (as in this scenario) an additional PCR can be sent at 2–3 weeks of age, and the results of the day 1 and week 2 to 3 PCR assays should be requested urgently. If the infant is found to be infected, then the baby should commence on HAART as soon as possible to reduce the risk of rapid HIV disease progression in the first year of life (Violari et al., 2008).

5.5 When Should Cotrimoxazole Prophylaxis be Started?

Before interventions to reduce MTCT were widely available, infection with *Pneumocystis jirovecii* was a common AIDS presentation in the first year of life, although very few cases occurred before 2 months of age (Simonds et al., 1995; Williams et al., 2001). Cotrimoxazole prophylaxis was shown to reduce the risk of *P. jirovecii* pneumonia (PCP) (Thea et al., 1996), and routine prophylaxis was previously recommended for all exposed infants starting at 4–6 weeks of age and continuing until their HIV infection status could be determined (CDC, 1995). The recommendation of 4–6 weeks treatment is based on the rarity of PCP before 2 months of age and the possibility that adverse effects will be more common in neonates, especially if taking zidovudine monotherapy or combination therapy (Chokephaibulkit et al., 2000; Bhana et al., 2002). With substantial reductions in MTCT and improvements in diagnosis of HIV infected infants by HIV DNA PCR, it is very unlikely that a low risk infant who has negative HIV DNA PCR tests at birth and 6 weeks will be HIV infected (Benjamin et al., 2001). If these results are available in a timely fashion and satisfactory follow-up can be ensured, then these infants will not require cotrimoxazole prophylaxis unless they are subsequently shown to be infected. For infants at higher risk of transmission, as in this clinical scenario, cotrimoxazole prophylaxis is recommended from 4 to 6 weeks of age until a third HIV DNA PCR at 3 months of age is confirmed to be negative. For infants with indeterminate status due to missed tests or conflicting test results, cotrimoxazole prophylaxis should be considered from 4 to 6 weeks and continued until HIV status is confirmed. For infants proven to be HIV infected, cotrimoxazole prophylaxis should be continued for the first year of life and then may be discontinued provided that the CD4+ T-cell count is satisfactory according to UK national guidelines (Chakraborty and Shingadia 2006).

5.6 What Is the Appropriate Management of a Mother Diagnosed with HIV Antenatally Who Has Refused Any Interventions?

Rarely, and usually for social or religious reasons, some women when diagnosed with HIV in pregnancy will deny their infection and refuse interventions. This is a challenging situation, and a successful outcome for the infant depends on the strenuous efforts of the multidisciplinary team to build a relationship with the woman before the delivery. Very often, over time, the woman can be encouraged to engage and to develop trust and confidence in the team and to finally undertake a package of appropriate care for herself and for her infant. This process often involves close liaison between the primary care team, the midwifery team, and social services to offer support where needed.

Where engagement does not occur, despite all the team's efforts, and the woman continues to refuse all interventions, then a pre-birth planning meeting with social services should be undertaken with the aim of safe guarding the child's future health. At this meeting, the woman should be informed that although she does not want treatment for herself, treatment should be made available to her infant at birth, and, at delivery, legal permission will be sought from a judge for the infant to be started as soon as possible on PEP and for breast feeding to be prevented. Sometimes this meeting will be sufficient to change the woman's view and engage her in treatment, but if not then legal intervention may be required (de Ruiter et al., 2008). Several such cases have been presented to the court over the last few years and infants have been successfully treated. Throughout these difficult situations all members of the team should try to remain supportive such that the woman can eventually accept her diagnosis, seek treatment for her own health, and remain well to look after her child.

6 Conclusions

The prevention of MTCT of HIV is the collective responsibility of many individuals and organisations involved in the provision of healthcare as well as the responsibility of the expectant mother. Changes in the way HIV testing is offered in pregnancy will be necessary to detect all women with HIV, including those seroconverting during pregnancy. Communication at all stages of management needs to be optimised so that there are no possibilities for system failures resulting in MTCT. Cases of transmission should be investigated to determine what went wrong as well as "near miss" cases of suboptimal management, as these will be far more frequent than actual transmissions. Paediatricians need to refer carefully to the BHIVA guidelines for all infants born to HIV infected mothers, and they need to be aware of any case where there is a potential increased risk of transmission so that expert advice can be sought and appropriate interventions can be undertaken.

References

Anonymous. Perinatal transmission of HIV in England 2002–2005. (2007). Audit Information and Analysis Unit for London, Kent, Surrey and Sussex, Essex, Bedfordshire and Hertfordshire. Executive summary available from http://www.chiva.org.uk/publications/PDF/2007/perinatal.pdf.

Antiretroviral Pregnancy Registry Steering Committee 2008. (2008). Antiretroviral Pregnancy Registry International Interim Report for 1 January 1989 through 31 July 2008. Wilmington, NC: Registry Coordinating Center. (Retrieved on 8/12/2008) www.APRegistry.com.

Benjamin, D.K., Jr., Miller, W.C., Fiscus, S.A., Benjamin, D.K., Morse, M., Valentine, M. et al. (2001). Rational testing of the HIV-exposed infant. *Pediatrics* (108), E3.

Bhana, N., Ormrod, D., Perry, C.M. & Figgitt, D.P. (2002). Zidovudine: a review of its use in the management of vertically-acquired pediatric HIV infection. *Paediatr Drugs* (4), 515–553.

Brahmbhatt, H., Sullivan, D., Kigozi, G., Askin, F., Wabwire-Mangenm, F., Serwadda, D., Sewankambo, N., et al. (2008). Association of HIV and malaria with mother-to-child transmission, birth outcomes, and child mortality. *J Acquir Immune Defic Syndr* (47), 472–476.

British Medical Association. Access to health care for asylum seekers and refused asylum seekers. (2008). Guidance from the BMA's medical ethics department. September 2008. (Retrieved on 8/12/2008) http://www.bma.org.uk/images/asylumhealthcare2008_tcm26-175519.pdf .

Bryson, Y.J., Luzuriaga, K., Sullivan, J.L. & Wara, D.W. (1992). Proposed definitions for in utero versus intrapartum transmission of HIV-1. *N Engl J Med* (327), 1246–1247.

Burns, D.N., Landesman, S., Muenz, L.R., Nugent, R.P., Goedert, J.J., Minkoff, H., Walsh, J.H., et al. (1994). Cigarette smoking, premature rupture of membranes, and vertical transmission of HIV-1 among women with low CD4+ levels. *J Acquir Immune Defic Syndr* (7), 718–726.

Capparelli, E.V., Mirochnick, M., Dankner, W.M., Blanchard, S., Mofenson, L., Mcsherry, G.D., et al. (2003). Pharmacokinetics and tolerance of zidovudine in preterm infants. *J Pediatr* (142), 47–52.

Centers for Disease Control. (1995). 1995 revised guidelines for prophylaxis against Pneumocystis carinii pneumonia for children infected with or perinatally exposed to human immunodeficiency virus. National Pediatric and Family HIV Resource Center and National Center for Infectious Diseases, Centers for Disease Control and Prevention. MMWR Recomm Rep, (44), 1–11.

Chakraborty, R. & Shingadia, D. (2006). Treating Opportunistic Infections In HIV-Infected Children (extended). Guidelines for the Children's HIV Association (CHIVA). 2006. (Retrieved on 08/12/2008) http://www.chiva.org.uk/protocols/ois-long.html#PCP.

Chokephaibulkit, K., Chuachoowong, R., Chotpitayasunondh, T., Chearskul, S., Vanprapar, N., Waranawat, N., Mock, P., et al. (2000). Evaluating a new strategy for prophylaxis to prevent Pneumocystis carinii pneumonia in HIV-exposed infants in Thailand. Bangkok Collaborative Perinatal HIV Transmission Study Group. *AIDS* (14), 1563–1569.

Coovadia, H.M., Rollins, N.C., Bland, R.M., Little, K., Coutsoudis, A., Bennish, M.L. & Newell, M.L. (2007). Mother-to-child transmission of HIV-1 infection during exclusive breastfeeding in the first 6 months of life: an intervention cohort study. *Lancet* (369), 1107–1116.

Coutsoudis, A., Dabis, F., Fawzi, W., Gaillard, P., Haverkamp, G., Harris, D.R., Jackson, J.B., et al. (2004). Late postnatal transmission of HIV-1 in breast-fed children: an individual patient data meta-analysis. *J Infect Dis* (189), 2154–2166.

De Ruiter, A., Mercey, D., Anderson, J., Chakraborty, R., Clayden, P., Foster, G., Gilling-Smith, C., et al. (2008). British HIV Association and Children's HIV Association guidelines for the management of HIV infection in pregnant women 2008. *HIV Med* (9), 452–502.

Dickover, R.E., Garratty, E.M., Herman, S.A., Sim, M.S., Plaeger, S., Boyer, P.J., Keller, M., et al. (1996). Identification of levels of maternal HIV-1 RNA associated with risk of perinatal transmission. Effect of maternal zidovudine treatment on viral load. *JAMA* (275), 599–605.

International Perinatal HIV Group. (2001). Duration of ruptured membranes and vertical transmission of HIV-1: a meta-analysis from 15 prospective cohort studies. *AIDS* (15), 357–368.

Foster, C.J. & Lyall, E.G. (2006). HIV in pregnancy: evolution of clinical practice in the UK. *Int J STD AIDS* (17), 660–667.

Hirt, D., Urien, S., Ekouévi, D.K., Rey, E., Arrivé, E., Blanche, S., Amani-Bosse, C. et al. (2009). Population pharmacokinetics of tenofovir in HIV-1-infected pregnant women and their neonates (ANRS 12109). *Clin Pharmacol Ther* (85), 182–189.

John, G.C., Nduati, R.W., Mbori-Ngacha, D.A., Richardson, B.A., Panteleeff, D., Mwatha, A., et al. (2001). Correlates of mother-to-child human immunodeficiency virus type 1 (HIV-1) transmission: association with maternal plasma HIV-1 RNA load, genital HIV-1 DNA shedding, and breast infections. *J Infect Dis* (183), 206–212.

Koulinska, I.N., Villamor, E., Chaplin, B., Msamanga, G., Fawzi, W., Renjifo, B. & Essex, M. (2006). Transmission of cell-free and cell-associated HIV-1 through breast-feeding. *J Acquir Immune Defic Syndr* (41), 93–99.

Kourtis, A.P., Butera, S., Ibegbu, C., Beled, L. & Duerr, A. (2003). Breast milk and HIV-1: vector of transmission or vehicle of protection? *Lancet Infect Dis* (3), 786–793.

Kuhn, L., Abrams, E.J., Matheson, P.B., Thomas, P.A., Lambert, G., Bamji, M., Greenberg, B., et al. (1997). Timing of maternal-infant HIV transmission: associations between intrapartum factors and early polymerase chain reaction results. New York City Perinatal HIV Transmission Collaborative Study Group. *AIDS* (11), 429–435.

Kwiek, J.J., Arney, L.A., Harawa, V., Pedersen, B., Mwapasa, V., Rogerson, S.J. & Meshnick, S.R.(2008). Maternal-fetal DNA admixture is associated with intrapartum mother-to-child transmission of HIV-1 in Blantyre, Malawi. *J Infect Dis* (197), 1378–1381.

Kwiek, J.J., Mwapasa, V., Milner, D.A., Jr., Alker, A.P., Miller, W.C., Tadesse, E., Molyneux, M.E., et al. (2006). Maternal-fetal microtransfusions and HIV-1 mother-to-child transmission in Malawi. *PLoS Med* (3), e10.

Landesman, S.H., Kalish, L.A., Burns, D.N., Minkoff, H., Fox, H.E., Zorrilla, C., Garcia, P., et al. (1996). Obstetrical factors and the transmission of human immunodeficiency virus type 1 from mother to child. The Women and Infants Transmission Study. *N Engl J Med* (334), 1617–1623.

Lehman, D.A. & Farquhar, C. (2007). Biological mechanisms of vertical human immunodeficiency virus (HIV-1) transmission. *Rev Med Virol* (17), 381–403.

Mackie, N.E., Fidler, S., Tamm, N., Clarke, J.R., Back, D., Weber, J.N. & Taylor, G.P. (2004). Clinical implications of stopping nevirapine-based antiretroviral therapy: relative pharmacokinetics and avoidance of drug resistance. *HIV Med* (5), 180–184.

Maiques, V., Garcia-Tejedor, A., Perales, A. & Navarro, C. (1999). Intrapartum fetal invasive procedures and perinatal transmission of HIV. *Eur J Obstet Gynecol Reprod Biol* (87), 63–67.

Mandelbrot, L., Mayaux, M.J., Bongain, A., Berrebi, A., Moudoub-Jeanpetit, Y., Benifla, J.L., et al. (1996). Obstetric factors and mother-to-child transmission of human immunodeficiency virus type 1: the French perinatal cohorts. SEROGEST French Pediatric HIV Infection Study Group. *Am J Obstet Gynecol* (175), 661–667.

Musoke, P., Guay, L.A., Bagenda, D., Mirochnick, M., Nakabiito, C., Fleming, T., Elliott, T. et al. (1999). A phase I/II study of the safety and pharmacokinetics of nevirapine in HIV-1-infected pregnant Ugandan women and their neonates (HIVNET 006). *AIDS* (13), 479–486.

Mwapasa, V., Rogerson, S.J., Kwiek, J.J., Wilson, P.E., Milner, D., Molyneux, M.E., et al. (2006). Maternal syphilis infection is associated with increased risk of mother-to-child transmission of HIV in Malawi. *AIDS* (20), 1869–1877.

Nesheim, S., Jamieson, D.J., Danner, S.P., Maupin, R., O'sullivan, M.J., Cohen, M.H., et al. (2007). Primary human immunodeficiency virus infection during pregnancy detected by repeat testing. *Am J Obstet Gynecol* (197), 149.e1–149.e5.

Newell, M.L. (1998). Mechanisms and timing of mother-to-child transmission of HIV-1. *AIDS* (12), 831–837.

Nielsen, K., Boyer, P., Dillon, M., Wafer, D., Wei, L.S., Garratty, E., et al. (1996). Presence of human immunodeficiency virus (HIV) type 1 and HIV-1-specific antibodies in cervicovaginal secretions of infected mothers and in the gastric aspirates of their infants. *J Infect Dis* (173), 1001–1004.

Pacifici, G.M. (2006). Pharmacokinetics of antivirals in neonate. *Early Hum Dev* (81), 773–780.

Peckham, C. & Gibb, D. (1995). Mother-to-child transmission of the human immunodeficiency virus. *N Engl J Med* (333), 298–302.

Prasitwattanaseree, S., Lallemant, M., Costagliola, D., Jourdain, G. & Mary, J.Y. (2004). Influence of mother and infant zidovudine treatment duration on the age at which HIV infection can be detected by polymerase chain reaction in infants. *Antivir Ther* (9), 179–185.

Rahangdale, L. & Cohan, D. (2008). Rapid human immunodeficiency virus testing on labor and delivery. *Obstet Gynecol* (112), 159–163.

Rousseau, C.M., Nduati, R.W., Richardson, B.A., Steele, M.S., John-Stewart, G.C., Mbori-Ngacha, D.A. et al. (2003). Longitudinal analysis of human immunodeficiency virus type 1 RNA in breast milk and of its relationship to infant infection and maternal disease. *J Infect Dis* (187), 741–747.

Simonds, R.J., Lindegren, M.L., Thomas, P., Hanson, D., Caldwell, B., Scott, G. & Rogers, M. (1995). Prophylaxis against Pneumocystis carinii pneumonia among children with perinatally acquired human immunodeficiency virus infection in the United States. Pneumocystis carinii Pneumonia Prophylaxis Evaluation Working Group. *N Engl J Med* (332), 786–790.

Thea, D.M., Lambert, G., Weedon, J., Matheson, P.B., Abrams, E.J., Bamji, M., Straus, W.L. et al. (1996). Benefit of primary prophylaxis before 18 months of age in reducing the incidence of Pneumocystis carinii pneumonia and early death in a cohort of 112 human immunodeficiency virus-infected infants. New York City Perinatal HIV Transmission Collaborative Study Group. *Pediatrics* (97), 59–64.

Thorne, C., Patel, D. & Newell, M.L. (2004). Increased risk of adverse pregnancy outcomes in HIV-infected women treated with highly active antiretroviral therapy in Europe. *AIDS* (18), 2337–2339.

Townsend, C.L., Cortina-Borja, M., Peckham, C.S., De Ruiter, A., Lyall, H. & Tookey, P.A. (2008a). Low rates of mother-to-child transmission of HIV following effective pregnancy interventions in the United Kingdom and Ireland, 2000–2006. *AIDS* (22), 973–981.

Townsend, C.L., Cortina-Borja, M., Peckham, C.S. & Tookey, P.A. (2007). Antiretroviral therapy and premature delivery in diagnosed HIV-infected women in the United Kingdom and Ireland. *AIDS* (21), 1019–1026.

Townsend, C.L., Cortina-Borja, M., Peckham, C.S. & Tookey, P.A. (2008b). Trends in management and outcome of pregnancies in HIV-infected women in the UK and Ireland, 1990–2006. *BJOG* (115), 1078–1086.

Tuomala, R.E., O'driscoll, P.T., Bremer, J.W., Jennings, C., Xu, C., Read, J.S., et al. (2003). Cell-associated genital tract virus and vertical transmission of human immunodeficiency virus type 1 in antiretroviral-experienced women. *J Infect Dis* (187), 375–384.

Tuomala, R.E., Shapiro, D.E., Mofenson, L.M., Bryson, Y., Culnane, M., Hughes, M.D., et al. (2002). Antiretroviral therapy during pregnancy and the risk of an adverse outcome. *N Engl J Med* (346), 1863–1870.

Violari, A., Cotton, M.F., Gibb, D.M., Babiker, A.G., Steyn, J., Madhi, S.A., et al. (2008). Early antiretroviral therapy and mortality among HIV-infected infants. *N Engl J Med* (359), 2233–2244.

Walmsley, S. (2003). Opt in or opt out: what is optimal for prenatal screening for HIV infection? *CMAJ* (168), 707–708.

Warszawski, J., Tubiana, R., Le Chenadec, J., Blanche, S., Teglas, J.P., Dollfus, C. et al. (2008). Mother-to-child HIV transmission despite antiretroviral therapy in the ANRS French Perinatal Cohort. *AIDS* (22), 289–299.

Watts, D.H., Balasubramanian, R., Maupin, R.T., Jr., Delke, I., Dorenbaum, A., Fiore, S. et al. (2004). Maternal toxicity and pregnancy complications in human immunodeficiency virus-infected women receiving antiretroviral therapy: PACTG 316. *Am J Obstet Gynecol* (190), 506–516.

Williams, A.J., Duong, T., Mcnally, L.M., Tookey, P.A., Masters, J., Miller, R., et al. (2001). Pneumocystis carinii pneumonia and cytomegalovirus infection in children with vertically acquired HIV infection. *AIDS* (15), 335–339.

The Use and Abuse of Antibiotics and the Development of Antibiotic Resistance

B. Keith English and Aditya H. Gaur

1 Introduction

The growing problem of antibiotic resistance has made the formerly routine therapy of many infectious diseases challenging, and, in rare cases, impossible. The widespread nature of the problem has led some experts to speculate about a "post-antibiotic era." Furthermore, though antibiotic resistance occurs in nature and is an inevitable consequence of even the most prudent antibiotic use, it is clear that our overuse and misuse of antibiotics is responsible for most of the recent increases in antibiotic resistance (McGowan, 1983; Austin et al., 1999; Arnold and Straus, 2005). Judicious and rational antibiotic use has the potential to limit the emergence of clinically important antibiotic resistance and may be able to reduce the impact of resistance that has already developed, effectively prolonging the shelf life of today's (and tomorrow's) antibiotics (Dowell et al., 1988). Nonetheless, the threat posed by antibiotic-resistant pathogens has been a wakeup call for modern medicine – developing new antibiotics is important but strategies to prevent infectious diseases (by immunization or other public health measures) will always be preferable when feasible.

In pediatrics, important recent developments in the area of antibiotic resistance have been the emergence of clinically relevant resistance to beta-lactam antibiotics (including cefotaxime and ceftriaxone) in *Streptococcus pneumoniae* (Bradley and Connor, 1991; Sloas et al., 1992; Adam, 2002; Klugman, 2002), and the emergence and spread of infections caused by community-acquired strains of methicillin-resistant *Staphylococcus aureus* (MRSA) (Herold et al., 1998; Hunt et al., 1999; Kaplan, 2006). In both pediatric and adult medicine, the development of resistance to multiple classes of antibiotics has transformed the lowly enterococcus from a "second-rate pathogen to a first-rate problem" (Barbara Murray) (Murray, 1990, 2000; Moellering, 1992; English, 2009).

B.K. English (✉)
The University of Tennessee Health Science Center, Memphis, Tennessee, USA
e-mail: kenglish56@me.com

A. Finn et al. (eds.), *Hot Topics in Infection and Immunity in Children VI*, Advances in Experimental Medicine and Biology 659, DOI 10.1007/978-1-4419-0981-7_6,
© Springer Science+Business Media, LLC 2010

2 Antibiotic Resistance Is Inevitable

In 1939, the microbiologist Rene Dubos discovered gramicidin, the first clinically tested antibiotic (it cured mice but proved too toxic for use as systemic treatment in humans, though topical gramicidin was used for many years) (Van Epps, 2006). However, Dubos' work with gramicidin encouraged Howard Florey at Oxford University to continue his work with penicillin, which had been discovered a decade before that by Alexander Fleming. Dubos' in vitro studies of these and other antimicrobial agents led to a landmark observation that was to prove to be of enormous clinical importance – that the exposure of bacteria to antibiotics over time often resulted in the development of antibiotic-resistant strains. Or, as Dubos put it in 1942, "susceptible bacterial species often give rise by 'training' to variants endowed with great resistance to these agents" (Moberg, 1996; Van Epps, 2006).

Ecological studies and studies of the bacterial flora of humans and animals living in areas with no exposure to antibiotics indicate that only low levels of antibiotic resistance are found in the absence of antibiotic exposure. For example, in the studies of Solomon Island natives, performed by Robert Moellering and colleagues, the microbial flora of these antibiotic-naïve individuals consisted largely of beta-lactam susceptible coliforms and streptococci (Gardner et al., 1969, 1970). Although the penicillin MICs of the enterococci isolated from these individuals were higher than those of the streptococcal isolates, the highest MIC observed in this group of isolates was 8 μg/mL (range 0.4–8 μg/mL) (Moellering et al., 1970). Studies of other similarly antibiotic-naïve human populations have provided comparable results (Davis and Anandan, 1970).

Ecologic studies of the bacterial flora of wild mammals have occasionally detected substantial rates of antimicrobial resistance in these animals, but these studies likely reflect exposure to humans and/or farm animals that have been treated with antibiotics. For example, while one group of researchers reported finding relatively high levels of antimicrobial resistance in commensal enterobacteria isolated from wild rodents in the British countryside (Gilliver et al, 1999), (suggesting that the development of resistance might be common in the absence of antibiotic exposure), other groups have failed to detect antibiotic resistance in *E. coli* isolated from wild animals in much more remote locations (e.g., northern Finland) where prior exposure to antimicrobials is extremely unlikely (Osterblad et al., 2001) (Fig. 1). Thus it seems likely that antimicrobial exposure ("pressure") is the prime mover behind the emergence of most clinically significant antimicrobial resistance.

Fig. 1 Antibiotic-resistant *E. coli* in animal feces

- 90% R to amoxicillin and cefuroxime (voles and wood mice, Wirral peninsula in NW England) (Gilliver et al, Nature 401:233, 1999)

- 0% R to amoxicillin and cefuroxime (voles, moose, white-tailed deer, two remote areas in Finland) (Osterblad et al, Nature 409: 38, 2001)

3 Antibiotic Overuse and Misuse Is a Major Problem

The evidence that antibiotics are overused and misused is incontrovertible (Dowell et al., 1998; Gaur and English, 2006). Misuse of antibiotics is inevitable in countries where antibiotics are available over the counter without a prescription (Gaur and English, 2006), but, clearly, the overuse and misuse of antibiotics is also a major problem in countries with the most sophisticated health care systems (Dowell et al., 1998; Arnold and Straus, 2005; Gaur and English, 2006).

Early discussions of antibiotic misuse in the United States and Europe often focused on the failure of the patient and family to facilitate "rational" antibiotic use, possibly because of the poor adherence to prescribed regimens and the stockpiling of antibiotics for later (usually inappropriate) use, etc. (Dowell et al., 1998). As guidelines promoting the "judicious" use of antibiotics by physicians have been promulgated, more recently, the focus has shifted to the prescriber, (Tomasz 1994; Dowell et al., 1998; Polk, 1999; Schwartz, 1999; Gaur and English, 2006). The judicious use of antibiotics requires more than a decision to prescribe or not prescribe an antimicrobial agent. The judicious health care provider uses an antibiotic only when indicated, chooses a cost-effective agent which provides appropriate antimicrobial coverage for the diagnosis that is suspected, and prescribes the optimal dose and duration of that antimicrobial based on the best available evidence (Gaur and English, 2006). The WHO Global Strategy for Containment of Antimicrobial Resistance defines the appropriate use of antimicrobials as the cost-effective use that maximizes clinical therapeutic effect while minimizing both drug-related toxicity and the development of antimicrobial resistance (http://www.who.int/drugresistance/WHO_Global_Strategy_English.pdf).

4 Antibiotic Use (and Abuse) Drives the Development of Resistance

Although antibiotic resistance may develop in the absence of antibiotic pressure, there is no doubt that antimicrobial usage drives the development of resistance in most cases (Tomasz, 1994; Dowell et al., 1998; Austin et al., 1999; Pichichero, 1999; Gaur and English, 2006).

Even optimal antibiotic use often leads to the development of resistance, and the diagnostic uncertainty of everyday clinical practice guarantees that some antibiotic "overuse" will occur even in optimal circumstances (Pichichero, 1999; Arnold et al., 2005). However, there is no question that the overuse and inappropriate use of antibiotics greatly contributes to the growing problem of antibiotic resistance (McGowan, 1983; Dowell et al., 1998; Arnold and Straus, 2005). In addition, efforts to promote both rational and judicious antibiotic use have clearly paid dividends in some settings. For example, a nationwide effort to reduce macrolide prescribing in Finland was associated with a decline in erythromycin resistance amongst group A streptococcal isolates (Seppala et al., 1997). Unfortunately, both mathematical models (Austin et al., 1999) and clinical experience (Belongia et al., 2001; Gaur and English, 2006) indicate that it takes much longer for reductions in antibiotic overuse to lead to reductions in antibiotic resistance than it takes for antibiotic overuse to trigger that resistance in the first place.

5 The Role of Antibiotic Use and Overuse in Facilitating the Emergence and Spread of Antibiotic-Resistant Pathogens Is Complex

While antibiotic exposure likely contributes to the development of all clinically-significant antibiotic resistance in bacteria, the effects are not uniform, and they vary with the organism and the mechanisms of antibiotic resistance. Ecological studies of staphylococci, pneumococci, and other bacteria indicate that resistance to particular antibiotics increases when the amount of that drug being used in that area increases (e.g., erythromycin resistance in *Staphylococcus aureus* in the 1950s and 1960s, beta-lactam resistance in the pneumococcus in the past two decades) (Arnold and Straus, 2005; Gaur and English, 2006). Widespread use of antibiotics in feed animals has been clearly associated with increasing rates of antimicrobial resistance in pathogens such as Salmonella and Campylobacter (Fey et al., 2000; Gorbach, 2001; White et al., 2001; Gerner-Smidt and Whichard, 2008). Use of the glycopeptide antibiotic avoparcin in animal feed in Europe has been associated with widespread human carriage of vancomycin-resistant enterococci. Similarly, the use of the streptogramin antibiotic virginiamycin in animal feed in the US has been associated with the emergence of enterococci resistant to the streptogramin combination quinupristin-dalfopristin in chickens and humans (McDonald et al., 2001).

In addition, the use of topical antibiotics in certain hospital settings has been strongly associated with the rapid emergence of resistance to the particular agent (e.g., development of gentamicin resistance in strains of *Pseudomonas aeruginosa* found in burns units) (reviewed by Gaur and English, 2006).

6 The Role of Antibiotic Use and Overuse in Facilitating the Emergence and Spread of Specific Antibiotic-Resistant Gram-Positive Pathogens in Pediatrics – Pneumococci, Staphylococci and Enterococci

6.1 Penicillin and Cephalosporin-Resistant Streptococcus pneumoniae

In the United States, clinically significant resistance to penicillins and cephalosporins in the pneumococcus was first clearly associated with cefotaxime/ceftriaxone treatment failure in children with bacterial meningitis caused by these strains (Bradley and Connor, 1991; Sloas et al., 1992). Since that time, beta-lactam resistance amongst pneumococci has become a major clinical problem and has been clearly been linked to treatment failure in both invasive (meningitis) and focal (otitis media) pneumococcal infections (McCracken, 1995; Adam, 2002; Schrag et al., 2004). It is likely that some patients with severe pneumococcal pneumonia also fail therapy with these cephalosporins (Feikin et al., 2000; English and Buckingham, 2006), but this appears to be rare and may only apply to infections caused by isolates with unusually high cefotaxime/ceftriaxone MICs. Less than 1% of pneumococci isolated from US patients with invasive disease have very-high-level penicillin resistance (MIC 8 μg/mL or higher), but larger percentages of isolates with high-level resistance to penicillin have been reported in some regions (e.g., 3.8% in Tennessee) (Schrag et al., 2004).

It seems likely that decades of exposure of *Streptococcus pneumoniae* to beta-lactam antibiotics have resulted in the sequential acquisition of multiple mutations in the peptidoglycan-assembling proteins that microbiologists refer to as "penicillin-binding proteins." These mutations eventually resulted in the emergence of clinically significant resistance to both penicillin and cephalosporin antibiotics.

In the community setting, many cross-sectional and case-control studies have shown that recent exposure to beta-lactam antibiotics is strongly associated with both nasopharyngeal carriage of resistant pneumococci and local or invasive disease caused by antibiotic-resistant pneumococci (reviewed by Arnold and Straus, 2005). These two phenomena – carriage and infection/disease – are of course closely related, as many previous studies have clearly indicated that pneumococcal infections are likely to develop within the first few weeks after acquisition of the carriage of a new strain. Thus antibiotic exposure appears to play important roles both in acquisition of pneumococcal carriage of and in pneumococcal disease caused by resistant organisms. However, the mechanism(s) that facilitate such colonization are poorly understood and may be different from those that promote such disease. For example, a recent serial cross-sectional study of nasopharyngeal pneumococcal carriage in children found that recent use of a cephalosporin by patients or siblings was associated with increased carriage of beta-lactam resistant pneumococci but had no effect on the carriage of susceptible pneumococci. Meanwhile, recent exposure to penicillins was associated with decreased carriage of susceptible pneumococcal isolates but not with increased carriage of resistant ones! (Samore et al., 2005).

6.2 Community-Associated, Methicillin-Resistant Staphylococcus aureus (CA-MRSA)

Methicillin-resistant strains of *Staphylococcus aureus* (MRSA) were first described in 1961 and Barrett and colleagues first reported hospital-associated infections caused by MRSA in the United States in 1968 (Barrett et al., 1968). Until recently, most serious infections caused by MRSA were nosocomial (Martin, 1994; Lowy, 1998), though community-acquired infections caused by MRSA have been reported since the 1970s. Recently, infections caused by community-associated strains of MRSA (CA-MRSA) have emerged as important causes of both minor and life-threatening infections in children and adults (Kaplan, 2006). Though first observed by Herold and colleagues in Chicago in the mid-1990s (Herold et al., 1998), the problem came to national attention in the United States in 1999 when the CDC reported four deaths resulting from CA-MRSA infections in children living in Minnesota and North Dakota (Hunt et al., 1999).

Though there is evidence that the use and misuse of antibiotics has facilitated the emergence and dissemination of CA-MRSA strains (Moran et al., 2006; Shorr, 2007), it is not clear that antibiotic pressure has played a major role in the rapid spread of dominant CA-MRSA strains such as those in the USA300 group. On the otherhand the use and misuse of antibiotics clearly played important roles in the previous emergence and spread of both penicillin resistance (in the 1950s) and nosocomial methicillin resistance (in the 1960s) in *Staphylococcus aureus* (Martin, 1994; Lowy, 1998). In fact, CA-MRSA infections emerged and spread rapidly in many communities at a time when a variety of efforts were being made (with variable degrees of success) to reduce unnecessary antibiotic use (Arnold and Straus, 2005).

6.3 Vancomycin-Resistant Enterococci (VRE)

The emergence of VRE as an important nosocomial pathogen (Schaberg et al., 1991; Moellering, 1992) has occurred almost entirely because of intrinsic and acquired resistance to antibiotics. Enterococci generally are organisms of low virulence, and antibiotic resistance has allowed them to persist both in the environment around and the gastrointestinal tracts of high-risk patients (mainly in the intensive care units of large teaching hospitals, in cancer centers, newborn ICUs, etc.) and to cause disease in immunocompromised patients who have received broad-spectrum antibiotics, which are effective against most other pathogens (and most "normal flora") (reviewed by English and Shenep, 2009).

Enterococci are intrinsically less susceptible to penicillin compared with streptococci and are "tolerant" to all beta-lactams. This intrinsic resistance and tolerance had a major clinical impact after the discovery of penicillin, as this "wonder drug" that cured many bacterial diseases (including almost all cases of endocarditis caused by viridans streptococci) failed to cure up to 2/3 of patients with enterococcal endocarditis (Hunter, 1947; Murray, 1990). Enterococci are intrinsically resistant to cephalosporins, and the widespread empiric use of broad-spectrum cephalosporins has played a major role in the emergence of VRE as a clinical problem (Murray, 2000; English and Shenep, 2009).

Furthermore, over the past four decades, enterococci have acquired high-level resistance to ampicillin, aminoglycosides, and glycopeptides. The overuse of antibiotics clearly contributed to the acquisition and persistence of these types of acquired resistance, but the effects of different types of antibiotics were not uniform. For example, many studies indicate that vancomycin overuse/misuse has played an important role in the emergence and spread of VRE, *but* recent data suggest that overuse of two other classes of antibiotics played a more important role. Exposure to broad-spectrum antibiotics that lack activity against enterococci (e.g., 3rd generation cephalosporins and carbapenems) has been associated with acquisition of VRE colonization. However, antibiotics that kill the anaerobic flora of the intestine promote high-grade *and* prolonged colonization with VRE (even if they have anti-enterococcal activity), likely playing a critical role in promoting invasive disease in immunocompromised patients (reviewed by English and Shenep, 2009).

7 Promoting Rational and Judicious Antibiotic Use; Is This a Pipe Dream?

In 1998, the Centers for Disease Control and the American Academy of Pediatrics argued that "Children can be protected from resistant bacteria through the judicious use of antimicrobial agents by their physicians" (Dowell et al., 1998). Can this "advocacy to protect our children from the spectre of the post-antibiotic era" really work or are we dreaming?

The answer to this question must come in two parts: First, will reducing unnecessary antibiotic use really help impede the emergence of resistance and/or reduce current rates of resistance? The answer to these questions is yes; however, the impact will vary by pathogen, by situation (e.g., hospital or community), and by antibiotic (Guillemot et al., 2005; Arnold and Straus, 2005). Importantly, preventing the emergence of resistance is always preferable to trying to reduce or eliminate it once it has occurred! (Austin et al., 1999; Belongia et al., 2001). Second, is it possible to identify strategies that can effectively change physician prescribing behavior? If we can, do we have the will (and the resources) to ensure that such strategies will be implemented? While efforts are underway to address the over-prescription of antibiotics by physicians (Arnold and Straus, 2005), we believe the answers to these critical questions remain uncertain.

As discussed above and at length elsewhere (Arnold and Straus, 2005; Gaur and English, 2006), the scope of the problem is vast. Examples of inappropriate use of antibiotics abound in the literature. For example, the landmark report by McCaig and Hughes (1995) found that there were almost 20 million antibiotic prescriptions for "upper respiratory infection" in the US in 1992. The factors accounting for the inappropriate use of antibiotics are not always clear, but recent efforts to understand this phenomenon are bearing fruit (Schwartz et al., 1997; Bauchner et al., 1999; Mangione-Smith et al., 1999; Hare et al., 2006; Arnold and Straus, 2005; Gaur and English, 2006).

One consistent result in these studies is that a physicians perception of what parents or patients desire often influences prescribing behavior – and those perceptions are often incorrect! (Mangione-Smith et al., 1999). Other misconceptions by physicians also influence prescribing behavior – e.g., the commonly-held, but probably incorrect, notion that it is more time-consuming to explain why an antibiotic is not indicated than to quickly prescribe one (Hare et al., 2006).

8 Summary, Conclusions, and Recommendations

There is clear evidence that antibiotic resistance develops under antibiotic pressure.While this may not be the only factor contributing to the development of antibiotic resistance and reduction in antibiotic use may not always be followed by a decrease in resistance, a decrease in antibiotic overuse will remain the number one intervention in our attempts towards slowing down the development of antimicrobial resistance. There are multitudes of patient and provider related factors that drive antibiotic use and overuse. While even the prudent, rational and judicious use of antibiotics can eventually lead to the emergence of antibiotic resistance in many cases, we can at least limit the development and spread of antibiotic resistance in clinical pratice by avoiding the unnecessary and inappropriate use of these important drugs. A number of interventions have been tried to promote judicious use of antibiotics around the world. many of them are discussed in the WHO Global strategy for containment of antimicrobial resistance (http://www.who.int/drugresistance/WHO_Global_Strategy_English.pdf). The applicability of these interventions differs not only based on the clinical setting of antibiotic use i.e. management of acute infections in an outpatient setting vs. inpatient setting vs. treatment of chronic infections, but also on a number of other factors such as site characteristics (private practise vs. academic setting), available resources (such as electronic data management and electronic prescriptions) and patient characteristics (literacy, cultural beliefs, socioeconomic status). No single intervention is likely to have a significant impact by itself and a combined approach using multiple interventions is necessary. Additionally, while the enormity of the problem and the degree to which it has become pervasive in society, especially in some countries maybe daunting, every effort that is made to promote judicious antibiotic use will have some benefit.

References

Adam, D. (2002). Global antibiotic resistance in *Streptococcus pneumoniae. J Antimicrob Chemother* 50(Suppl), 1–5.

Arnold, S.R. & Straus, S.E. (2005). Interventions to improve antibiotic prescribing practices in ambulatory care. *Cochrane Database Syst Rev* (19), CD003539 1–65.

Arnold, S.R. et al. (2005). Antibiotic prescribing for upper respiratory tract infection: the importance of diagnostic uncertainty. *J Pediatr* (146), 222–226.

Austin, D.J., Kristinsson, K.G., & Anderson, R.M. (1999). The relationship between the volume of antimicrobial consumption in human communities and the frequency of resistance. *Proc Natl Acad Sci USA* (96), 1152–1156.

Bauchner, H., Pelton, S.I., & Klein J.O. (1999). Parents, physicians, and antibiotic use. *Pediatrics* (103), 395–401.

Barrett, F.F., McGehee, R.F., Jr., & Finland, M. (1968). Methicillin-resistant *Staphylococcus aureus* at Boston City Hospital. Bacteriologic and epidemiologic observations. *N Engl J Med* (279), 441–448.

Belongia, E.A., Sullivan, B.J., Chyou, P-H., Madagame, E., Reed, K.D., & Schwartz, B. (2001). A community intervention trial to promote judicious antibiotic use and reduce penicillin-resistant *Streptococcus pneumoniae* carriage in children. *Pediatrics* (108), 575–583.

Bradley, J.S. & Connor, J.D. (1991). Ceftriaxone failure in meningitis caused by *Streptococcus pneumoniae* with reduced susceptibility to beta-lactam antibiotics. *Pediatr Infect Dis J*, (10), 871–873.

Davis, C.E. & Anandan, J. (1970). The evolution of r factor: a study of a preantibiotic community in Borneo. *N Engl J Med* (282), 117–122.

Dowell S.F., Marcy S.M., Phillips, W.R., Gerber M.A., & Schwartz, B (1998). Principles of judicious use of antimicrobial agents for pediatric upper respiratory tract infections. *Pediatrics* 163–165.

English, B.K. & Buckingham, S.C. (2006). Impact of antimicrobial resistance on therapy of bacterial pneumonia in children. *Adv Exp Med Biol* (582), 125–135.

English, B.K. & Shenep, J.L. (2009). Enterococcal and viridans streptococcal infections. In: R.D. Feigin, J.D. Cherry, G. Demmler & S.L. Kaplan (Eds.) *Textbook of Pediatric Infectious Diseases* (6th Ed.). Philadelphia: Saunders pp. 1258–1288.

Feikin, D.R., Schuchat, A., Kolczak, M., Barrett, N.L., Harrison, L.H., Lefkowitz, L., McGeer, A. et al. (2000). Mortality from invasive pneumococcal pneumonia in the era of antibiotic resistance, 1995–1997. *Am J Public Health* (90), 223–229.

Fey, P.D., Safranek, T.J., Rupp, M.E., Dunne, E.F., Ribot, E., Iwen, P.C., Bradford, P.A, Angulo, F.J. et al. (2000). Ceftriaxone-resistant salmonella infection acquired by a child from cattle. *N Engl J Med* (342), 1242–1249.

Gardner, P., Smith, D.H., Beer, H., & Moellering, R. (1969). Recovery of resistance (R) factors from a drug-free community. *Lancet* 2(7624), 774–776.

Gardner, P., Smith, D.H., Beer, H., & Moellering R.C. Jr. (1970). Recovery of resistance factors from a drug-free community. *Lancet* 1(7641), 301.

Gaur, A.H., & English, B.K. (2006). The judicious use of antibiotics – an investment towards optimized health care. *Indian J Pediatr* (73), 343–350.

Gerner-Smidt, P., & Whichard, J.M. (2008). Foodborne disease trends and reports. *Foodborne Pathog Dis* (5), 551–554.

Gilliver, M.A., Bennett, M., Begon, M., Hazel, S.M., & Hart, C.A. (1999). Antibiotic resistance in wild rodents. *Nature* (401), 233–234.

Gorbach, S.L. (2001). Antimicrobial use in animal feed – time to stop. *N Engl J Med* (345), 1202–1203.

Guillemot, D., Varon, E., Bernede, C., Weber, P., Henriet, L., Simon, S., Laurent, C. et al. (2005). Reduction of antibiotic use in the community reduces the rate of colonization with penicillin G-nonsusceptible *Streptococcus pneumoniae*. *Clin Infect Dis* (41), 930–938.

Hare, M.E., Gaur, A.H., Somes, G.W., Arnold, S.R., & Shorr, R.I. (2006). Does it really take longer not to prescribe antibiotics for viral respiratory tract infections in children? *Ambul Pediatr* (6), 152–156.

Herold, B.C., Immergluck, L.C., Maranan, M.C., Lauderdale, D.S., Gaskin, R.E., Boyle-Vavra, S., Leitch, C.D. et al. (1998). Community-acquired methicillin-resistant *Staphylococcus aureus* in children with no identified predisposing risk. *JAMA* (279), 593–598.

Hunt, C., Dionne, M., Delorme, M. et al. (1999). Four pediatric deaths from community-acquired methicillin-resistant *Staphylococcus aureus*: Minnesota and North Dakota, 1997–1999. *MMWR* (48), 707–710.

Hunter, T.H. (1947). Use of streptomycin in treatment of bacterial endocarditis. *Am J Med*, (2), 436–442.

Kaplan, S.L. (2006). Community-acquired methicillin-resistant *Staphylococcus aureus* infections in children. *Semin Pediatr Infect Dis* (17), 113–119.

Klugman, K.P. (2002). The successful clone: the vector of dissemination of resistance in Streptococcus pneumoniae. *J Antimicrob Chemother* 50 (S2): 1–5.

Lowy, FD. (1998). Medical progress: *Staphylococcus aureus* infections. *N Engl J Med* (339), 520–532.

Mangione-Smith, R., McGlynn, E.A., Elliott, M.N., Krogstad, P., & Brook, R.H. (1999). The relationship between perceived parental expectations and pediatrician antimicrobial prescribing behavior. *Pediatrics* (103), 711–718.

Martin, MA. (1994). Methicillin-resistant *Staphylococcus aureus*: the persistent resistant nosocomial pathogen. *Curr Clin Top Infect Dis* (14), 170–191.

McCaig, L.F., & Hughes, JM. (1995). Trends in antimicrobial drug prescribing among office-based physicians in the United States. *JAMA* (274), 214–219.

McCracken, G.H., Jr. (1995). Emergence of resistant *Streptococcus pneumoniae*: a problem in pediatrics. *Pediatr Infect Dis J* (14), 424–428.

McDonald L.C., Rossiter S, Mackinson C., Wang Y.Y., Johnson S., Sullivan M., Sokolow R. et al. (2001). Quinupristin-dalfopristin-resistant Enterococcus faecium on chicken and in human stool specimens. *N Engl J Med* (345), 1155–1160.

McGowan, J.E. Jr. (1983). Antimicrobial resistance in hospital organisms and its relation to antibiotic use. *Rev Infect Dis* (5), 286–291.

Moberg, C.L. (1996). Rene Dubos: a harbinger of microbial resistance to antibiotics. *Microb Drug Resist* (2), 287–297.

Moellering, R.C., Jr., Wennersten, C., & Medrek, T. (1970). Prevalence of high-level resistance to aminoglycosides in clinical isolates of enterococci. *Antimicrob Agents Chemother* (1), 335–340.

Moellering, R.C., Jr. (1992). Emergence of *Enterococcus* as a significant pathogen. *Clin Infect Dis* (14), 1173–1176.

Moran, G.J., Krishnadasan A., Gorwitz R.J., Fosheim G.E., McDougal L.K., Carey R.B., Talan D.A. (2006). EMERGEncy ID Net Study Group. Methicillin-resistant *S. aureus* infections among patients in the emergency department. *N Engl J Med* (355), 666–674.

Murray, B.E. (1990). The life and times of the enterococcus. *Clin Microbiol Rev* (3), 46–65.

Murray, B.E. (2000). Vancomycin-resistant enterococcal infections. *N Engl J Med* (342), 710–721.

Osterblad, M., Norrdahl, K., Korpimaki, E., & Huovinen, P. (2001). How wild are wild mammals? *Nature* (409), 37–38.

Pichichero, M.E. (1999). Understanding antibiotic overuse for respiratory tract infections in children. *Pediatrics* (104), 1384–1388.

Polk, R. (1999). Optimal use of modern antibiotics: emerging trends. *Clin Infect Dis* (29), 264–274.

Samore, M.H., Lipsitch, M., Alder, S.C., Haddadin, B., Stoddard, G., Williamson, J., Sebastian, K. et al. (2005). Mechanisms by which antibiotics promote dissemination of resistant pneumococci in human populations. *Amer J Epidem* (163), 160–170.

Schaberg, D.R., Culver, D.H., & Gaynes, R.P. (1991). Major trends in the microbial etiology of nosocomial infection. *Am J Med* 91(Suppl 3B), 72–75.

Schrag, S.J., McGee, L., Whitney, C.G., Beall, B., Craig, A.S., Choate, M.E., Jorgensen, J.H. et al. (2004). Active Bacterial Core Surveillance Team. 2004. Emergence of *Streptococcus pneumoniae* with very-high-level resistance to penicillin. *Antimicrob Agents Chemother* (48), 3016–3022.

Schwartz, B., Bell, D.M., & Hughes, J.M. (1997). Preventing the emergence of antimicrobial resistance. A call for action by clinicians, public health officials, and patients. *JAMA* (278), 944–945.

Schwartz, B. (1999). Preventing the spread of antimicrobial resistance among bacterial respiratory pathogens in industrialized countries: the case for judicious antimicrobial use. *Clin Infect Dis* (28), 211–213.

Seppala, H., Klaukka, T., Vuopio-Varkila, J., Muotiala, A, Helenius, H., Lager, K., & Huovinen, P. (1997). The effect of changes in the consumption of macrolide antibiotics on erythromycin resistance in group A streptococci in Finland. Finnish Study Group for Antimicrobial Resistance. *N Engl J Med* (337), 441–446.

Shorr, A.F. (2007). Epidemiology of staphylococcal resistance. *Clin Infect Dis* (45), S171–S176.

Sloas, M.M., Barrett, F.F., Chesney, P.J., English, B.K., Hill, B.C., Tenover, F.C., & Leggiadro, R.J. (1992) Cephalosporin treatment failure in penicillin- and cephalosporin-resistant *Streptococcus pneumoniae* meningitis. *Pediatr Infect Dis J* (11), 662–666.

Tomasz, A. (1994). Multiple-antibiotic-resistant pathogenic bacteria. A report on the Rockefeller University Workshop. *N Engl J Med* (330), 1247–12451.

Van Epps, HL. (2006). Rene Dubos – unearthing antibiotics. *J Exp Med* (203), 259.

White, D.G., Zhao, S., Sudler, R., Ayers, S., Friedman, S., Chen, S. et al. (2001). The isolation of antibiotic-resistant salmonella from retail ground meats. *N Engl J Med* (345), 1147–1154.

Vaccination Against Varicella: What's the Point?

Anne A. Gershon

1 Introduction

Live attenuated varicella vaccine was developed in Japan in the early 1970s as a means to prevent varicella primary infection with varicella-zoster virus (VZV) in healthy and immunocompromised individuals (Takahashi et al., 1974). Takahashi had attenuated the virus by passage in various cell cultures using different propagation temperatures and produced a candidate vaccine virus that proved to be both safe and immunogenic. Initially, there was great controversy as to whether it was likely to be safe to develop a vaccine that had the potential to become latent after immunization and that might even be oncogenic (Brunell, 1977, 1978). It was also questioned as to whether varicella was a serious enough disease to be worth preventing.

Despite numerous objections, during the 1980s, a number of clinical trials in leukemic and healthy varicella-susceptible children and adults were undertaken, culminating in 1995, in the licensure of varicella vaccine and its recommendation as a routine childhood immunisation in the United States (Gershon et al., 2008). Clinical trials indicated that the Takahashi vaccine strain Oka was highly immunogenic, safe, and produced long-term immunity in healthy children and selected immunocompromised children. By the turn of the century, most of the published data on varicella vaccine had emanated from the United States, following commencement of routine administration (Centers for Disease Control, 1996). This manuscript discusses recent data on the saga of live attenuated varicella vaccine use in healthy children following its licensure in the United States.

By way of background, varicella is a generalized illness with a moderately long incubation period (2 weeks) and viremic phases as the rash develops. It represents primary infection with VZV. During varicella, latent infection with VZV develops in dorsal root ganglia (DRG) and cranial nerve ganglia (CNG). For the lifetime of most individuals the virus remains quiescent, but about 30% go on

A.A. Gershon (✉)
Department of Pediatrics, Columbia University College of Physicians and Surgeons, New York, NY 10032, USA
e-mail: aag1@columbia.edu

A. Finn et al. (eds.), *Hot Topics in Infection and Immunity in Children VI*, Advances in Experimental Medicine and Biology 659, DOI 10.1007/978-1-4419-0981-7_7, © Springer Science+Business Media, LLC 2010

to manifest clinical zoster, usually in the 6th to the 8th decade of life when cell mediated immunity (CMI) to VZV wanes due to normal aging (Oxman et al., 2005). Zoster is a secondary infection with VZV, due to the same virus that was acquired during varicella (Breuer, 2003). Simply stated, varicella develops in persons with no humoral immunity to VZV; zoster develops in persons lacking CMI to VZV.

The Oka strain of live attenuated varicella virus was developed in the 1970s by Takahashi and his colleagues in Osaka, Japan (Takahashi et al., 1974). It was subjected to small studies there, and there seemed to be interest in bringing the vaccine to the west. In 1978, an international conference was held at NIH that included outstanding basic and clinical virologists of the era, to discuss whether varicella vaccine should be studied in the United States. A plethora of clinical trials in leukemic children and healthy adults and children soon followed, and, in 1995, as a result of their success regarding safety, immunogenicity, and protection, the vaccine was licensed in the United States (Gershon et al., 2008, 1984). One dose was recommended for children less than 13 years old, and two doses (at least 1 month apart) was recommended for older susceptible individuals. Routine immunisation was begun for children between the ages of 12 and 18 months.

At present, there is no question that this vaccine is highly effective. The incidence of varicella in the vaccinated and unvaccinated has decreased by about 80% as have hospitalizations for chickenpox and mortality from them (Gershon et al., 2008; Seward et al., 2002). A case control study indicated that not only was the vaccine about 85% protective from clinical varicella but also, with tine over 8 years after vaccination, there was no loss of immunity. (Gershon et al., 2007; Vazquez et al., 2001; 2004). While an epidemiologic study suggested that there might be waning immunity after varicella vaccination, it did not consider the potential importance of primary vaccine failure that might be confused with secondary vaccine failure (Gershon et al., 2007; Chaves et al., 2007). A number of other studies also indicated that the vaccine was about 85% effective in preventing varicella (Gershon et al., 2008). However, it was disconcerting, that despite high rates of vaccine coverage, outbreaks of chickenpox still occurred in schools. When vaccine efficacy was calculated during these outbreaks, it was reported to vary from 44 to 100%, with an average of about 80%. The vaccine was not proving to be as effective as it had been projected from clinical trials (Gershon et al., 2008).

An additional potential concern about the vaccine was whether there would be an increase in the incidence of varicella in young adults who were not vaccinated, who had never achieved immunity after vaccination, or who had became only temporarily immune to varicella. Another concern was whether there would be a serious increase in the incidence of zoster in individuals who had experienced natural varicella in the past but lost CMI due to a decrease in re-exposure to VZV because vaccination resulted in decreased circulation of the virus.

Understanding basic aspects of the natural history of VZV infection is helpful in understanding the pathogenesis of both varicella and zoster. VZV is a highly cell-associated virus when propagated in cell culture. It does not emerge in infectious form from the cells in which it grows, and therefore its only means of spread is from one cell to another (Chen et al., 2004). In the human body, VZV spreads by the same means, slowly from one cell to another. This slow spread from cell to cell explains why the humoral antibody response has little impact on recovery from varicella; cell free virions, which can be neutralized by specific antibodies, do not develop in the body as VZV spreads. In order to defend against VZV that is spreading in the body, CMI is required. However, there is one place in the body where cell-free virus is released from infected cells, and this is in the superficial epidermis, where one of the receptors of VZV that confines it to cells is not synthesized; as a result, VZV is able to escape from the cell in infectious form. VZV in vesicle fluid is cell free, well formed, and highly infectious; it is also capable of causing latent infection. Because latent virus can remain safe for decades in a host and later emerge and infect a whole new population of younger individuals who have never previously seen this virus and are therefore susceptible to it, this has enormous significance for the biology and survival of VZV. In turn, these younger individuals, develop chickenpox followed by latency, and the virus survives for another generation.

There is evidence that preventing VZV from infecting the skin can decrease the incidence of zoster. Children with underlying leukemia have a high incidence of zoster during their chemotherapeutic treatment if they have had varicella. In contrast, vaccinated leukemic children have a lower incidence of development of zoster, although those with a history of a vaccine associated VZV rash are more likely to develop zoster than vaccinees with no rash (Hardy et al., 1991). Hypothesis exist that there is less zoster because most vaccinated children do not develop VZV rashes. In vitro studies of the Oka strain of VZV indicate that it is quite capable of causing latent infection with VZV (Chen et al., 2003). It seems likely that varicella vaccine is not only effective in preventing varicella but also in preventing zoster. The lower incidence of zoster in vaccinees as compared to those with previous natural infection has been found in leukemic children following renal transplantation and in children with HIV infection (Gershon et al., 2008). Whether this will also be true for healthy children will not be known for certain for decades when these children, vaccinated in the mid-1990s, become older adults.

How can we explain the failure of 15% of vaccinated children to develop protective immunity to varicella after vaccination? This level of protection is lower than that seen after measles and rubella immunization. Could it be due to the fact that VZV causes latent infection? Is there another factor that is involved, and what might this be?

At one time, it was thought that the dose of virus in a live vaccine was unimportant as long as the virus could multiply in the host in order to stimulate the immune system. However, studies with measles vaccine failed to confirm this hypothesis (Aaby et al., 1996). We now know that generally speaking for varicella vaccine, the higher the immunizing dose the stronger the immune response (Krause and Klinman, 1995). Nevertheless, when varicella vaccine was being developed, it was common to determine the lowest dose that would stimulate an antibody response. Possibly the dose of virus in the licensed vaccine in the US is on the low side for stimulation of immunity to VZV. Although clinical trials suggested that this dose stimulated humoral immunity in about 95% of children, the test that was used yielded false positive results when low levels of immunity were being assessed (Provost et al., 1989, 1991). In contrast, a group of 138 children in New York, Tennessee, and California were tested for seroconversion after receiving 1 dose of the vaccine licensed in the United States, using the fluorescent antibody to membrane antigen (FAMA) assay, which is the "gold standard" (Williams et al., 1974); only 76% of these children seroconverted (Michalik et al., 2008). These results were one of the reasons that a second dose of varicella vaccine was mandated in 2006 by the Centers for Disease Control and Prevention (CDC) for all children. At present, studies in the US are underway to determine whether the second dose confers more protection than only 1 dose.

It is projected that 2 doses will provide more protection than 1 dose, and that use of 2 doses of measles-mumps-rubella-varicella vaccine will be even more effective against varicella than two doses of the monovalent vaccine (Kuter et al., 2004; Shinefield et al., 2006). The goal is to decrease circulation of VZV to the lowest possible levels and to have a highly immunized population in order to decrease both varicella and zoster. Judicious use of the zoster vaccine may also help to achieve this goal (Oxman et al., 2005).

References

Aaby, P., Samb, B., Simondon, F. et al. (1996). A comparison of vaccine efficacy and mortality during routine use of high-titre Edmonston-Zagreb and Schwarz standard measles vaccines in rural Senegal. *Trans R Soc Trop Med Hyg*, (90), 326–330.

Breuer, J. (2003). Monitoring virus strain variation following infection with VZV: is there a need and what are the implications of introducing the Oka vaccine? *Commun Dis Public Health*, (6), 59–62.

Brunell, P.A. (1977). Commentary: protection against varicella. *Pediatrics*, (59), 1–2.

Brunell, P.A. (1978). Varicella vaccine: the crossroads is where we are not! *Pediatrics*, (62), 858–859.

Centers-for-Disease-Control. (1996). Prevention of varicella: Recommendations of the Advisory Committee on Immunization Practices (ACIP). *Morb Mort Wkly Rep*, (45), 1–36.

Chaves, S.S., Gargiullo, P., Zhang, J.X. et al. (2007). Loss of vaccine-induced immunity to varicella over time. *N Engl J Med*, (356), 1121–1129.

Chen, J., Gershon, A., Silverstein, S.J., Li, Z.S., Lungu, O., & Gershon, M.D. (2003). Latent and lytic infection of isolated guinea pig enteric and dorsal root ganglia by varicella zoster virus. *J Med Virol*, (70), S71–S78.

Chen, J.J., Zhu, Z., Gershon, A.A., & Gershon, M.D. (2004). Mannose 6-phosphate receptor dependence of varicella zoster virus infection in vitro and in the epidermis during varicella and zoster. *Cell*, (119), 915–926.

Gershon, A.A., Arvin, A.M., & Shapiro, E. (2007). Varicella vaccine. *N Engl J Med*, (356), 2648–2649.

Gershon, A.A., Steinberg, S., & Gelb, L. (1984). NIAID-Collaborative-Varicella-Vaccine-Study-Group. Live attenuated varicella vaccine: efficacy for children with leukemia in remission. *JAMA*, (252), 355–362.

Gershon, A., Takahashi, M., & Seward, J. (2008). Live attenuated varicella vaccine. In: S. Plotkin, W. Orenstein, & P. Offit (Eds.) *Vaccines*. pp. 915–958, Philadelphia: WB Saunders.

Hardy, I.B., Gershon, A., Steinberg, S., LaRussa, P. et al. (1991). The incidence of zoster after immunization with live attenuated varicella vaccine. A study in children with leukemia. *N Engl J Med*, (325), 1545–1550.

Krause, P. & Klinman, D.M. (1995). Efficacy, immunogenicity, safety, and use of live attenuated chickenpox vaccine. *J Pediatr*, (127), 518–525.

Kuter, B., Matthews, H., Shinefield, H. et al. (2004). Ten year follow-up of healthy children who received one or two injections of varicella vaccine. *Pediatr Infect Dis J*, (23), 132–137.

Michalik, D.E., Steinberg, S.P., LaRussa, P.S. et al. (2008). Primary vaccine failure after 1 dose of varicella vaccine in healthy children. *J Infect Dis*, (197), 944–949.

Oxman, M.N., Levin, M.J., Johnson, G.R. et al. (2005). A vaccine to prevent herpes zoster and postherpetic neuralgia in older adults. *N Engl J Med*, (352), 2271–2284.

Provost, P.J., Krah, D.L., Kuter, B.J. et al. (1991). Antibody assays suitable for assessing immune responses to live varicella vaccine. *Vaccine*, (9), 111–116.

Provost, P., Krah, D., Miller, W. et al. (1989). Comparative sensitivities of assays for antibodies induced by live varicella vaccine. Interscience Conference on Antimicrobial Agents and Chemotherapy.

Seward, J.F., Watson, B.M., Peterson, C.L. et al. (2002). Varicella disease after introduction of varicella vaccine in the United States, 1995–2000. *JAMA*, (287), 606–611.

Shinefield, H., Black, S., Thear, M. et al. (2006). Safety and immunogenicity of a measles, mumps, rubella and varicella vaccine given with combined Haemophilus influenzae type b conjugate/hepatitis B vaccines and combined diphtheria-tetanus-acellular pertussis vaccines. *Pediatr Infect Dis J*, (25), 287–292.

Takahashi, M., Otsuka, T., Okuno, Y., Asano, Y., Yazaki, T., Isomura, S. (1974). Live vaccine used to prevent the spread of varicella in children in hospital. *Lancet*, (2), 1288–1290.

Vazquez, M., LaRussa, P., Gershon, A., Steinberg, S., Freudigman, K., & Shapiro, E. (2001). The effectiveness of the varicella vaccine in clinical practice. *N Engl J Med*, (344), 955–960.

Vazquez, M., LaRussa, P.S., Gershon, A.A. et al. (2004). Effectiveness over time of varicella vaccine. *JAMA*, (291), 851–855.

Williams, V., Gershon, A., & Brunell, P. (1974). Serologic response to varicella-zoster membrane antigens measured by indirect immunofluorescence. *J Infect Dis*, (130), 669–672.

What Can We Learn from the Retina in Severe Malaria?

Simon J. Glover, Kondwani Kawaza, Yamikani Chimalizeni, and Malcolm E. Molyneux

1 Introduction

Plasmodium falciparum malaria can cause death through a variety of complications, an important one of these being cerebral malaria (CM). In areas where *P. falciparum* transmission is intense, deaths from CM tend to occur in children under the age of 8 years – older people being protected by partial immunity. In an endemic area, asymptomatic *P. falciparum* parasitaemia is common. Therefore, among individuals with any clinical syndrome, a considerable proportion may have parasitaemia that is incidental and not causally related to the illness (Koram and Molyneux, 2007). This frequently leads to diagnostic difficulty. Children presenting with encephalopathy (altered consciousness with or without convulsions), who are parasitaemic may be suffering from cerebral malaria or may have another cause of encephalopathy (e.g. viral or metabolic) with incidental parasitaemia.

In view of the clinical diagnostic and, therefore, management dilemmas surrounding cerebral malaria, studies have focussed on finding features that occur specifically in cerebral malaria that may help distinguish CM from other causes of coma.

Malarial retinopathy is a recently described collection of signs that can be identified with the ophthalmoscope in patients with CM, but not in patients with encephalopathy of other causes. Being specifically associated with malaria, the presence of the distinctive retinopathy in a patient can add confidence to a clinical diagnosis of CM. This may be of value for clinical management, for designing entry and endpoint criteria in clinical trials, and for describing the disease burden attributable to malaria in a population.

M.E. Molyneux (✉)
College of Medicine, University of Malawi, Blantyre, Malawi; Malawi-Liverpool-Wellcome Trust Clinical Research Programme, Blantyre, Malawi
e-mail: mmolyneux000@googlemail.com

A. Finn et al. (eds.), *Hot Topics in Infection and Immunity in Children VI*, Advances in Experimental Medicine and Biology 659, DOI 10.1007/978-1-4419-0981-7_8,
© Springer Science+Business Media, LLC 2010

2 Malarial Retinopathy – The Components

Malarial retinopathy is unique to severe malaria and is commonly found in children with CM. Malarial retinopathy consists of a collection of signs, some of which are individually pathognomonic. A retina affected by malaria may have one or any combination of the following components: retinal whitening; hemorrhages; vessel changes, and swelling of the optic disc.

Retinal whitening consists of irregular, patchy areas that may be localized or may be widely distributed in all segments of the retina. The color of affected retina varies from subtle pallor to dense white. The pattern of retinal whitening in CM is distinctive in its distribution, usually affecting both the central macula (Fig. 1) and peripheral retina, although macular and peripheral whitening can occur independently of each other. There is often whitening along the temporal raphe (Fig. 2), i.e. on the same horizontal level as the fovea, but just lateral to the macula. (This is a watershed area for the retinal circulation).

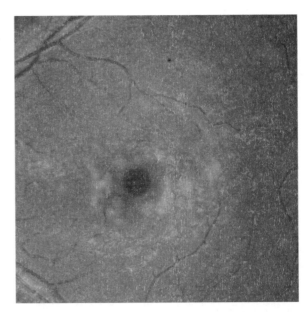

Fig. 1 Macular whitening centered around the fovea, the center of the fovea is the *central circular dark area* on this photograph

Hemorrhages are typically white-centered (Fig. 2) and can be of any size, position and number in the fundus. In a comatose child with *P. falciparum* parasitaemia, white-centered hemorrhages are highly suggestive of CM, although they may occur with other conditions, such as subacute bacterial endocarditis. Anemia can cause retinal hemorrhages, and patients with CM are often anaemic. However, hemorrhages are less commonly found and are fewer in number, in patients with severe malarial anemia (SMA) than in patients with CM (Beare et al., 2004a).

Vessel changes are pathognomonic of severe malaria. Large and medium-sized vessels may appear white or orange, while capillaries look white. Changes in larger

Fig. 2 Macular whitening, whitening along the temporal raphe, and white-centered retinal hemorrhages (at the inferior *left*-hand corner of the picture)

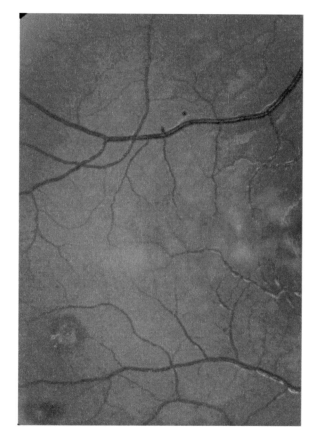

vessels are commonly segmental (Fig. 3), affecting variable lengths of scattered arterioles and venules, but capillary whitening can affect large areas of the fundus, often co-localizing with retinal whitening.

Optic disc swelling is not unique to CM and may occur in a variety of conditions that can cause raised intracranial pressure. The presence of papilloedema in the absence of other features of malarial retinopathy in a comatose patient increases the likelihood that there is a cause of raised intracranial pressure other than malaria.

It is important to differentiate clinically between optic disc swelling and optic disc hyperemia, the latter not being associated with high intracranial pressure. Optic disc hyperemia may be seen in various conditions, including meningitis and in CM. It may be a conspicuous feature, with the disc turning bright red, but its distinguishing feature is that the disc remains flat.

Retinal signs in survivors resolve over – one to four weeks, with no visual sequelae identifiable using currently available techniques (Beare et al., 2004a, 2004b). In endemic areas CM is uncommon in adults, but when it occurs, malaria retinopathy may be observed (Beare et al., 2003).

Fig. 3 White segment in a retinal blood vessel

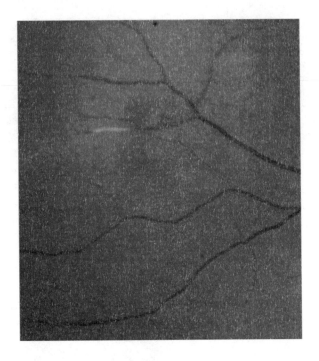

3 How to See Malarial Retinopathy

As with any retinal condition, malarial retinopathy is seen better after the pupil has been dilated with eye drops.

Atropine should not be used, because its action is prolonged and there is a theoretical risk of inducing amblyopia in a child who recovers quickly. Cyclopentolate is often used in pediatric eye clinics, as it has a much shorter duration of action than atropine and also has the useful secondary effect of causing cycloplegia, thus allowing more accurate testing for glasses in children. This is not necessary for diagnostic examinations. A useful degree of brief, reversible pupil dilatation can be achieved using a combination of *tropicamide* and *phenylephrine*. The African-pigmented iris does not respond to mydriatic drops as well as paler irides and therefore, tropicamide alone does not usually give enough mydriasis to be useful. Tropicamide 1%, in combination with phenylephrine, 2.5% is usually effective.

It is important to remember to examine the pupils for size and reactions before applying the eye drops, as this cannot be done afterwards for a few hours. Clearly mark in the patient's notes that mydriatic drops have been given. The disadvantage of temporarily losing the pupil reactions for diagnostic or monitoring purposes is usually compensated for by the useful diagnostic information gained. But if pupillary reactions are critical for the assessment of your patient over the next few hours, you may have to avoid applying mydriatic drops.

Most clinicians are familiar with using the direct ophthalmoscope. With this instrument, central retinal signs are easier to locate and identify than peripheral changes. Most cases of malarial retinopathy can be diagnosed in this fashion, as the central retina is affected in the majority of cases. Indirect ophthalmoscopy provides a fuller and quicker view of the retina and allows a more accurate assessment of disc swelling. Ideally, both methods should be used, but many pediatricians and general physicians do not have the facilities or the training to make use of indirect ophthalmoscopy.

4 The Prognostic Value of Retinopathy

In CM, the presence and severity of malarial retinopathy is related to length of coma in survivors and risk of death (Beare et al., 2004a; Lewallen et al., 1996). In a child with clinically defined CM, the presence of any malarial retinopathy carries a relative risk for death of 3.7 (95% confidence interval 1.6–8.5). Retinopathy is present in a significantly higher proportion of those who die than in survivors ($p = 0.001$). Increasing severity of retinal signs is related to increasing risk of a fatal outcome ($p < 0.05$). In survivors, retinal signs are associated with prolonged time to recovery of consciousness ($p < 0.001$). Papilloedema has the greatest relative risk of death, followed by vessel changes, retinal hemorrhages, peripheral retinal whitening and macular whitening (Beare et al., 2004a).

5 Malarial Retinopathy in Clinical Practice and Research

The retinopathy described above appears to be specific to severe malaria. Children with proven bacterial meningitis, viral meningoencephalitis or other causes of coma do not have these retinal signs. In an autopsy study of children dying with clinically diagnosed CM (unrousable coma, parasitemia and no other evident cause of the illness), post-mortem examination revealed that 24 of 31 had a pathological diagnosis of CM – i.e. there was sequestration of *P. falciparum* parasites in cerebral vessels and no alternative cause of death was found. The remaining seven had few or no *P. falciparum* parasites in brain microvessels and, in these, a likely alternative cause of death was identified. In that study, retinopathy (recorded during the illness) was found to be better than any other clinical or laboratory feature in distinguishing patients with CM from those with other pathologically identified diagnoses (Taylor et al., 2004).

For the clinician managing a patient with coma and parasitemia, antimalarial-specific and supportive treatment must be given, whether there is retinopathy or not, and alternative or additional diagnoses must be considered, irrespective of the presence or absence of retinal changes. But finding malarial retinopathy greatly increases the likelihood that CM is the correct diagnosis.

Malarial retinopathy has great potential for use in research studies of the pathogenesis or treatment of CM. If some of the patients enrolled to such studies do not have CM, the accuracy of findings will be compromised. In large-scale studies of anti-malarial interventions, the incidence of CM in trial populations may be used as an endpoint in the assessment of efficacy or effectiveness. Again, the capacity to identify CM accurately, using retinopathy as a confirmation of the diagnosis, may enhance the accuracy of the study's findings.

6 Retinal Angiography in CM

The injection of fluorescein into a peripheral vein makes it possible to obtain photographs of the retinal vascular tree, and in particular to study the contours, patency and integrity of retinal vessels.

(a) Vessel contours. In many patients with malarial retinopathy, vessel contours appear normal, but, in some, it is possible to detect filling defects and irregularity of vessel walls (Fig. 4). It seems likely that these abnormalities are due to the presence of sequestered parasites and sometimes microthrombi, as seen on retinal and cerebral histology in fatal cases.

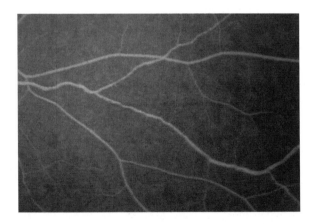

Fig. 4 Irregularity of vessel walls and intravascular filling defects – retinal angiogram

(b) Vessel patency. In some patients with retinopathy, some vessel segments are shown by angiography to be entirely occluded, with no passage of fluorescein. Blocked vessels coincide with vessels or segments that appear white on ophthalmoscopy. Mapping of unperfused areas has shown that they coincide precisely with areas of retinal whitening visible at ophthalmoscopy (Figs. 5 and 6) (Beare et al., 2009).

(c) Vessel integrity. Angiography sometimes shows leakage of fluorescein out of the intravascular compartment into the surrounding retina.

Fig. 5 Non perfusion in the *center* of the macula and at its periphery. The isolated area of non perfusion at the periphery of the macula and the periphery of this picture (*upper left*), is in exactly the same position as an area of retinal whitening shown in Fig. 6

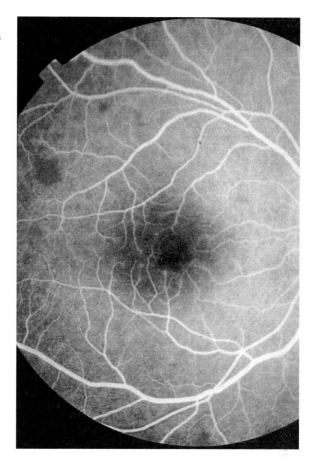

7 Pathogenesis of Retinal Findings and Potential Relevance to the Pathogenesis of Cerebral Disease

In embryogenesis, the retina develops as an extension of the diencephalon, and retinal and cerebral vessels share a similar developmental pattern. The retinal microvasculature has a blood-retinal barrier similar to the blood-brain barrier (Patton et al., 2005). In fatal CM, histopathological changes in the retina resemble those in the brain, including sequestration of parasitized erythrocytes, microthrombi and petechial hemorrhages (Lewallen et al., 2000).

Analysis of ophthalmoscopic appearances, angiographic findings and (in a few fatal cases) retinal histopathology has provided some likely explanations for the characteristic changes of malarial retinopathy. The orange or white discoloration of vessels is probably due to the sequestration of parasitized erythrocytes containing little residual hemoglobin, and each incorporating a granule of hemozoin, the by-product of hemoglobin that has been digested by a *P. falciparum* trophozoite. Some

Fig. 6 The retinal whitening at the *center* of this picture is in exactly the same place as the area of non-perfusion of the peripheral macula shown in Fig. 5. There is also an area of dense whitening inferior and to the *left* of the fovea, which matches the unusually large area of peri-foveal non perfusion also shown in Fig. 5

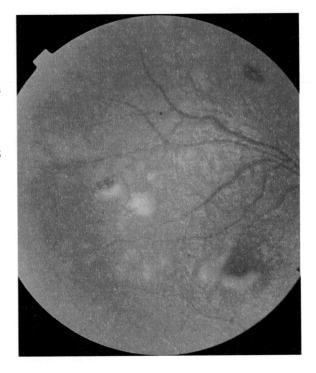

affected vessels are shown by angiography to be occluded, while others are patent. Retinal whitening co-localizes with areas of angiographic non-perfusion, and therefore probably results from retinal hypoxia or ischaemia. Microhemorrhages occur at sites of capillary rupture, sometimes, but not always, accompanied by the presence of a microthrombus in the affected vessel. In patients who have undergone both ophthalmoscopy during their illness and subsequent autopsy, the numbers of retinal and intracerebral hemorrhages correlate closely (White et al., 2001).

8 Summary

A characteristic retinopathy can be found by ophthalmoscopy in patients with cerebral malaria (CM). The presence of this retinopathy enhances the confidence with which a diagnosis of CM can be made. Fundoscopy can improve accurate enrollment of cases for studies of malaria pathogenesis or treatment, and can strengthen identification of CM as an endpoint in preventive interventions against malaria. Retinal changes may be instructive about pathological changes that are otherwise unobservable within the brain, and the retina offers the advantage that it can be directly observed during life. With the addition of angiography, retinal studies suggest that CNS microvessels are commonly occluded in CM, and also that the blood-retinal barrier is sometimes breached; an observation that is in keeping with the cerebral edema commonly identified at autopsy in fatal pediatric malaria.

References

Beare, N.A.V., Lewis, D.K., Kublin, J.G., Harding, S.P., Zijlstra, E.E., & Molyneux, M.E. (2003). Retinal changes in adults with cerebral malaria. *Ann Trop Med Parasitol,* (97), 313–315.

Beare, N.A., Southern, C., Chalira, C., Taylor, T.E., Molyneux, M.E., & Harding, S.P. (2004a). Prognostic significance and course of retinopathy in children with severe malaria. *Arch Ophthalmol,* (122), 1141–1147.

Beare, N.A., Southern, C., Kayira, K., Taylor, T.E., Harding, S.P. (2004b). Visual outcomes in children in Malawi following retinopathy of severe malaria. *Br J Ophthalmol,* (88), 321–324.

Beare, N.A.V., Harding, S.P., Taylor, T.E., Lewallen, S., & Molyneux, M.E. (2009). Perfusion abnormalities in children with cerebral malaria and malarial retinopathy. *J Infect Dis,* 199(2), 263–271.

Koram, K. & Molyneux, M.E. (2007). When is "malaria" malaria? The different burdens of malaria infection, malaria disease and malaria-like illnesses. In: J.G.Breman, M.S.Alilio, N.J.White (Eds.) Defining and defeating the intolerable burden of malaria – III. Progress and perspectives. *Am J Trop Med Hyg,* 77(6) supplement, 1–327.

Lewallen, S., Bakker, H., Taylor, T.E., Wills, B.A., Courtright, P., & Molyneux, M.E. (1996). Retinal findings predictive of outcome in cerebral malaria. *Trans R Soc Trop Med Hyg,* (90), 144–146.

Lewallen, S., White, V.A., Whitten, R.O., Gardiner, J., Hoar, B., Lindley, J., Lochhead, J., McCormick, A., Wade, K., Tembo, M., Mwenechanya, J., Molyneux, M.E., & Taylor, T.E. (2000). Clinical-histopathological correlation of the abnormal retinal vessels in cerebral malaria. *Arch Ophthalmol,* (118), 924–928.

Patton, N., Aslam, T., MacGillivray, T., Pattie, A., Deary, I.J., & Dhillon, B. (2005). Retinal vascular image analysis as a potential screening tool for cerebrovascular disease: a rationale based on homology between cerebral and retinal microvasculatures. *J Anat,* (206), 319–348.

Taylor, T.E., Fu, W.J., Carr, R.A., Whitten, R.O., Mueller, J.G., Fosiko, N.G., Lewallen, S., Liomba, N.G., & Molyneux, M.E. (2004). Differentiating the pathologies of cerebral malaria by postmortem parasite counts. *Nat Med,* (10), 143–145.

White, V.A., Lewallen, S., Beare, N., Kayira, K., Carr, R.A., & Taylor, T.E. (2001). Correlation of retinal hemorrhages with brain hemorrhages in children dying of cerebral malaria in Malawi. *Trans R Soc Trop Med Hyg,* (95), 618–621.

Epidemiology and Prevention of Neonatal Candidiasis: Fluconazole for All Neonates?

David A. Kaufman

1 Scope of the Problem

The epidemiology and prevention of invasive *Candida* infections (ICI) depends largely on risk factors and the role they play in infection control and identifying high-risk patients for prevention. Prevention of ICI is critical, as there is significant mortality and even with successful eradication, neurodevelopmental impairment occurs in 57% of survivors weighing less than1000 g (Stoll et al., 2004; Benjamin et al., 2006). It is important to know local invasive fungal infection data in your unit, as the incidence in available literature lacks important key information regarding practices that may increase or decrease the incidence. This includes resuscitation practices than may limit the number of infants below 25 weeks gestational age, surgical subpopulations (necrotizing enterocolitis, focal bowel perforation, gastroschisis), use of antifungals, and feeding practices. Single- and multi-center studies as well as meta-analyses, including the Cochrane review, have demonstrated significant efficacy with fluconazole prophylaxis in preterm infants (Kaufman, 2008c; Clerihew et al., 2007). Antifungal prophylaxis with nystatin still needs further study in extremely preterm infants. With local infection data and evidenced-based prevention, ICI can be nearly eliminated in every NICU.

2 Risk Factors for ICI

The major risk factors for ICI are outlined in Table 1. *Candida* can colonize, proliferate and disseminate from many sites: skin, gastrointestinal tract, respiratory tract, and central venous catheters. Specifically, yeasts adhere to catheters, organizing as biofilms. Facilitating this process, parenteral nutrition and lipid emulsions serve as growth media. Antibiotic and medication practices can increase infection as broad

D.A. Kaufman (✉)
Division of Neonatology, Department of Pediatrics, University of Virginia Health System, Charlottesville, VA 22908, USA
e-mail: dak4r@virginia.edu

A. Finn et al. (eds.), *Hot Topics in Infection and Immunity in Children VI*, Advances in Experimental Medicine and Biology 659, DOI 10.1007/978-1-4419-0981-7_9, © Springer Science+Business Media, LLC 2010

Table 1 Risk factors

In preterm infants
Extreme prematurity
Colonization
Central venous catheter
Prior bacterial bloodstream infection
Antibiotic and Medication Practices
Enteral feedings practices
Duration of mechanical ventilation
(Chronic Lung Disease ± postnatal steroids)
In any age neonate
Complicated gastrointestinal (GI) disease

spectrum antibiotics, such as third- and fourth-generation cephalosporins and carbapenems, can achieve high concentrations in the bile and significantly decrease the gram-negative and anaerobic competitive microflora of the gastrointestinal tract, enabling fungal colonization, proliferation and dissemination. The acidity of the stomach helps control the burden of yeast colonization, and H_2 antagonists are associated with increased risk for infection, as well necrotizing enterocolitis (Stoll et al., 1999; Cotton et al., 2006; Guillet et al., 2006). The immunomodulation by postnatal steroids also plays a role in increasing risk for infections (Stoll et al., 1999; Makhoul et al., 2007). In addition, BPD only and BPD with postnatal-steroid therapy are independently associated with an increased risk for *Candida* BSI (Makhoul et al., 2007). Factors that may decrease infection include breast milk, probiotics, lactoferrin, feeding practices and antifungal prophylaxis, which will be discussed further below (Manzoni et al., 2006d; Hylander et al., 1998).

Changes to the microflora of the skin, gastrointestinal and respiratory tracts play a significant role in the pathogenesis of ICI (Kaufman et al., 2001; 2006; Manzoni et al., 2006; Manzoni et al., 2006; Saiman et al., 2001). Prior bacterial BSI, exposure to 2 or more antibiotics, broad spectrum antibiotics as mentioned above (third generation cephalosporins, carbapenems) lack of enteral feedings, ileus, blood pressure or respiratory instability, or gastrointestinal disease, endotracheal tube intubation and duration of mechanical ventilation have been associated with increased risk for colonization or infection (Benjamin et al., 2006; Makhoul et al., 2007; Saiman et al., 2000; Benjamin et al., 2003; Feja et al., 2005). While these risk factors are important, the majority of colonization occurs in the first two weeks of life and antifungal prophylaxis and prevention are significantly more effective when started in the first 48 hours of life (Kaufman et al., 2001; 2005; Manzoni et al., 2007b; Ozturk et al., 2006). Prophylaxis is most effective when started early, as it prevents infection by inhibiting colonization (skin, gastrointestinal, respiratory, and central venous catheter) and multisite colonization (Kaufman et al., 2001; Manzoni et al., 2006). In infants weighing less than 1000 g on the third day of life, the following were associated with increased or decreased incidence of candidemia or *Candida* meningitis: lack of enteral feeding (8.7%) compared to receiving enteral feeding (3.4%), and exposure to third generation cephalosporins (15.3%) compared to other antibiotics (5.6%) or no antibiotics (0.9%) (Benjamin et al., 2006).

3 Incidence of ICI

Several studies have examined ICI in neonates reporting a varying incidence between centers and a change of incidence over time. In the largest analysis, data from the National Nosocomial Surveillance System Hospitals (NNIS) from 1995 to 2004 [132 neonatal intensive care units (NICUs) and 130,523 neonates] indicated that for infants weighing less than 1000 g at birth, 50% of NICUs had fungal bloodstream infection rates of greater than 7.5% and 25% had rates greater than 13.5% (Table 2) (Fridkin et al., 2006). However, there was considerable variation between NICUs ranging from 3 to 23% for the 10th and 90th percentiles. During a shorter time period (1998–2001) a smaller multi-center study reported similar results when examining BSI and meningitis (Cotton et al., 2006). Rates may differ due largely to the demographics of admitted patients, resuscitation practices, surgical population, feeding and antibiotic practices; unfortunately one or more of these factors are lacking from all epidemiologic studies to date. This is exemplified in two recent studies in the United Kingdom (UK). While overall rates of ICI in infants weighing less than 1000 g were reported to be 2.1% in the UK, antifungal prophylaxis was not ascertained for that study, but in a follow up survey almost 30% of UK NICUs used antifungal prophylaxis (Clerihew et al., 2006; Clerihew and McGuire, 2007). These were also the larger NICUs and possibly represent a higher percentage of the UK's extremely preterm infants. Therefore, due to the use of antifungal prophylaxis, the rates of infection in the UK may be lower than in the United States and other studies.

Table 2 Incidence of *Candida* bloodstream infections (%) by birth weight groups[23]

Percentiles	3rd	10th	50th	75th	90th
<1000 g	3	4.6	7.5	13.5	23
1000–1500 g	0	0	0.7	2.7	5.6
1501–2500 g	0	0	0	0.5	1.1
>2500 g	0	0	0	0.4	0.8

4 ICI Variation Among NICUs

Most studies only report BSIs, and fail to account for meningitis, urinary tract infections (UTIs), of which one-third have renal abscess involvement, and sterile body fluids such as peritoneal infections complicating NEC and focal bowel perforations. Weitkamp et al. recently reported the importance of examining ICI incidence in each NICU, as many flaws exist when relying solely on literature for the condition (Weitkamp et al., 2008). This NICU reported that while their rate of *Candida* BSIs was 6.8%, the rate of all ICI was 10% for infants weighing less than 1000 g.

Gestational age has a more linear relationship to ICI compared to birth weight and it aids in defining the highest risk patients (Makhoul et al., 2007; Clerihew et al., 2006; Johnsson and Ewald, 2004; Hack, 2006). For example, growth charts demonstrate that a 24-week gestation infant could be between 468 and 940 g (3rd to 97th percentiles) (Kramer et al., 2001). Examining the incidence of ICI by each gestational age as well as by birth weight (by each 100 g), NICUs should be able to see where their rates fall to zero. Table 3 is an example of an infectious control approach for each NICU to analyze its incidence of ICI with infection-related mortality and neurodevelopmental impairment.

Table 3 Invasive *Candida* infection (ICI) surveillance chart

Gestational age	All ICI (%)	Mortality	NDI	Bloodstream infections (%)	UTI (%)	Meningitis (%)	Other[a] (%)
22							
23							
24							
25							
26							
27							
28							
29							
30							

ICI, Invasive *Candida* Infections: Bloodstream, Urinary tract infections (UTIs), Meningitis, and [a]Other infections (Peritoneal and/or other sterile body fluid). NDI, Neurodevelopmental impairment (one or more of the following: psychomotor or mental developmental index of less than 70; Cerebral Palsy; Blindness; Deafness)

5 *Candida* Species and ICI

The majority (90%) of ICI are due to *C. albicans* and *C. parapsilosis* (Benjamin et al., 2006; Clerihew et al., 2006; Burwell et al., 2006). While the incidence of *C. parapsilosis* has increased over the past 15 years, *C. albicans* still comprises more than 50% of cases. From the NNIS study discussed above, *Candida* BSI distribution was as follows: *C. albicans* 58%, *C. parapsilosis* 34%, *C. tropicalis* 4%, *C. glabrata* 2%, *C. lusitaniae* 2%, and *C. krusei* 0.2%. Some NICUs have reported higher rates of *C. glabrata* (Parikh et al., 2007; Fairchild et al., 2002). These data are important in choosing the correct antifungal for treatment as well as prophylaxis.

6 Neurodevelopmental Impairment

In a multi-center analysis by the National Institute of Child Health and Human Development (NICHD) in infants weighing less than 1000 g, neurodevelopmental impairment occurred in 57% of survivors with *Candida* BSI and 53% with meningitis (Benjamin et al., 2006). Compared to other infections, *Candida* BSIs have the highest associated neurodevelopmental impairment (Figure 1) (Stoll et al., 2004).

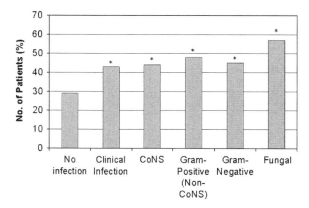

Fig. 1 Neurodevelopmental Impairment and Bloodstream Infection in Infants Weighing Less Than 1000 g.* P less than 0.001 compared to no infection group. Neurodevelopmental impairment: one or more of the following: psychomotor or mental developmental index of less than 70; Cerebral Palsy; visual or hearing impairment. Clinical Infection: late-onset infection with negative cultures with antibiotic treatment for more than 5 days. CoNS: Coagulase-negative staphylococci

7 Mortality

In infants weighing less than 1000 g, with ICI, mortality rates range from 23 to 66% of the control patients in the prophylaxis studies (Kaufman et al., 2001; Manzoni et al., 2006; Kaufman et al., 2005; Weitkamp et al., 2008; Kicklighter et al., 2001; Healy et al., 2005; Bertini et al., 2005; Manzoni et al., 2007; Uko et al., 2006; Aghai et al., 2006; Healy et al., 2008). In epidemiologic studies, mortality has been reported in several different ways, including overall, attributable, early (within 3–7 days of diagnosis), and infection-related, with rates ranging from 13 to 26% (Benjamin et al., 2004; 2005; Stoll et al., 2002; Makhoul et al., 2005; Zaoutis et al., 2007).

Variability in the incidence of mortality has also been noted with some single-center studies reporting no mortality in preterm infants. These differences may be due to standardization of diagnosis, initial antifungal dosing, empiric antifungal use and prompt removal of central catheters. Additionally, there is a marked difference in mortality between infants weighing less than 1000 g and larger infants with ICI in multi-center studies. A recent analysis using ICD-9 codes reported an overall mortality rate in infants weighing less than 1000 g of 26%, compared to 13% without candidiasis and for infants weighing more than 1000 g with ICI, mortality was 2% compared to 0.4% among infants without candidiasis (Fridkin et al., 2006; Zaoutis et al., 2007; Smith et al., 2007).

8 Costs of ICI

In additional to the morbidity and mortality, two recent case-controlled studies have examined the effect of ICI on cost of hospitalization and length of hospital stay (Zaoutis et al., 2007; Smith et al., 2007). The mean increase in hospital costs was $39,045 for infants weighing less than 1000 g with no difference in length of stay, and for infants weighing 1000 g or more, there was an increase of $122,302 with an additional length of stay of 16 days (Zaoutis et al., 2007).

9 Prompt and Empiric Treatment of ICI

Several studies have advocated empiric antifungal therapy in high-risk patients, with a resulting improvement in mortality in infants weighing less than 1500 g (Procianoy et al., 2006; Makhoul et al., 2002); however, in a multi-center study, for infants weighing less than 1000 g prompt or empiric therapy did not decrease neurodevelopmental impairment or mortality (Benjamin et al., 2006). Therefore, while prompt and empiric treatment is important, prevention should be paramount in each NICU to eliminate ICI as a cause of neurodevelopmental impairment or mortality.

10 Prophylaxis

10.1 Fluconazole

Fluconazole prophylaxis targets all sites of potential colonization and dissemination, e.g. skin, gastrointestinal tract, respiratory tract and intravenous catheters – central or peripheral – when given intravenously, via a central venous catheter, if present (Kaufman et al., 2001). The mechanisms of action as a prophylaxis agent are summarized in Table 4. Fluconazole prophylaxis may achieve even greater efficacy and safety in neonates compared with pediatric and adult immunocompromised patients due to the ability to start prophylaxis shortly after birth on day of life one (prior to colonization and proliferation) and the need for a much shorter duration of prophylaxis (4–6 weeks in neonates compared to 75–100 days in bone marrow transplant patients).

Table 4 Fluconazole prophylaxis and the prevention of invasive *Candida* infections

---

Pharmacokinetics (Walsh et al., 2000; Vaden et al., 1997; Koks et al., 2001)
- High tissue and body fluid concentrations (compared to plasma concentrations)
  1. Skin levels are 10–40 times greater
  2. Mucosal levels are 120–140%
  3. Urine concentrations are 10 times greater
  4. CSF penetration is 70–90%

- Long half-life
  1. Persistent Mucocutaneous Concentrations
  2. Skin and mucosal concentrations last 3–7 days after one dose

- Low lipophilicity
- Low protein binding
- 80% is excreted unchanged in the urine.
- These characteristics allow for long dosing intervals, excellent tissue penetration, easy elimination, with lower than treatment doses to achieve prevention

Animal and in vitro DataBU (Kojic and Darouiche, 2004; Ellepola and Samaranayake, 1998; Vaden et al., 1997; Hazen et al., 2000; Ghannoum et al., 1992; Houang et al., 1990; Schuman et al., 1997; Faergemann and Laufen, 1993)

Candida killing
- Fungal eradication
- Render fungi more susceptible to host defenses[a]
  o Even at sub therapeutic concentrations

Inhibits biofilms
- Inhibition of biofilm formation with central catheters

Impairs fungal adherence
- *Candida* adhesion is impaired
  o Epithelium
  o Endothelium (even at sub therapeutic concentrations)

In Preterm Infants: (Kaufman et al., 2001; Manzoni et al., 2006a; Manzoni et al., 2006b, Kicklighter et al., 2001; Manzoni et al., 2007a; Manzoni et al., 2007b)
Prevention and reduction of colonization

- Prevents colonization
  o Skin
  o Respiratory tract
  o Gastrointestinal tract
  o Central venous catheter
- Limits colonization/proliferation
  o Decreases number of sites colonized

---

[a]Enhanced phagocytosis and monocyte killing.

The efficacy and safety of fluconazole prophylaxis in preterm infants has been reported in more than 2600 patients from four randomized, controlled trials (Kaufman et al., 2001; 2005; Kicklighter et al., 2001; Manzoni et al., 2007b) and seven retrospective studies (Manzoni et al., 2006; Weitkamp et al., 2008; Healy et al., 2005; Bertini et al., 2005; Uko et al., 2006; Aghai et al., 2006; Healy et al., 2008) without any significant adverse effects or emergence of resistance (Table 5) (Kaufman, 2008a; Kaufman, 2008c).

Table 5 Fluconazole prophylaxis studies and invasive *Candida* infections (2001–2007)

Study	Enrollment criteria	Protocol	N	Dosing schedule	Fluconazole-treated	Placebo group	P
Randomized controlled trials (RCT)							
Kaufman et al. (2001)	<1000 g ETT or CVC ≤5 d of life	Single center Placebo-control	100	3 mg/kg IV × 6 wks or less when IV access no longer needed Schedule A	0 of 50 (0%)[a]	10 of 50 (20%)[a] 8 of 50 (16%)[c] *4 Candida-related deaths*	0.008 0.007[c]
Manzoni et al. (2007)	<1500 g <3 days of life	Multi-center Placebo-control Three-arm study	322	6 mg/kg or 3 mg/kg IV/PO 1000–1500 g: × 30 d <1000 g: × 45 d Schedule D	7 of 216 (3.2%)[a] *No Candida-related deaths*	14 of 106 (13.2%)[a] *2 Candida-related deaths*	<0.0001
Kaufman et al. (2005)	<1000 g ETT or CVC ≤5 d of life	Single center comparison of two dosing schedules	81	3 mg/kg IV × 6 wks or less when IV access no longer needed Schedule A vs. B	2 of 41 (5%)[a c] vs. 1 of 40 (2.5%)[a c] *No Candida-related deaths*	Dose comparison study. No placebo group	0.68
Kicklighter et al. (2001)	<1500 g <72 h of life	Single center Placebo-control Colonization study (1° endpoint)	103	6 mg/kg IV/PO Schedule C	1 of 53(1.9%)[a c] *1 Candida-related death*	0 of 50 (0%)[a]	0.52
Retrospective studies with historic controls							
Healy et al. (2005)	<1000 g ETT or CVC ≤5 d of life	Retrospective historic controls (2000–2001) vs. FP patients (2002–2004)	446	3 mg/kg IV × 6 wks or less when IV access no longer needed Schedule A	3 of 240 (1.3%)[b] *No Candida-related deaths*	15 of 206 (7%)[b] *4 Candida-related deaths*	0.01

Table 5 (continued)

Study	Enrollment criteria	Protocol	N	Dosing schedule	Fluconazole-treated	Placebo group	P
Aghai et al. (2006)	<1000 g	Retrospective historic controls (1998–2002) vs. FP patients (2002–2005)	277	3 mg/kg IV × 6 wks or less when IV access no longer needed Schedule A	0 of 140 (0%)	9 of 137 (6.6)[c] 6 Candida-related deaths	0.006
Manzoni et al. (2006)	<1500 g	Retrospective historic controls (1998–2000) vs. FP patients (2001–2003)	465	6 mg/kg IV/PO 1000–1500 g: × 30 d, <1000 g: × 45 d Schedule D	4 of 225 (1.8%)[a] No Candida-related deaths	22 of 240 (9.2%)[a] 4 Candida-related deaths	0.001
Bertini et al. (2005)	<1500 g & CVC	Retrospective historic controls (1998–2000) vs. FP patients (2001–2003)	255	6 mg/kg IV/PO × 4 wks Schedule C	0 of 136 (0%)[a]	9 of 119 (7.6%)[a c] 3 Candida-related deaths	0.003
Uko et al. (2006)	<1500 g or <32 wks Only during antibiotic administration of >3 days	Retrospective historic controls (2001–2002) vs. FP patients (2003–2004)	384	3 mg/kg IV <30 wks Schedule E ≤30 weeks Schedule F	2 of 178 (1.1%)[d] No Candida-related deaths	13 of 206 (6.3%)[a c] 2 Candida related deaths in infants<1000 g	0.007

FP, fluconazole prophylaxis; IV, intravenous; PO, enteral; CVC, central venous catheter; RCT, Randomized Controlled Trials.

[a] Bloodstream, Urine and CSF (cerebrospinal fluid) infections during FP.

[b] Bloodstream and CSF infections during FP.

[c] Number of bloodstream infections.

[d] Only 91 of the 178 patients received FP. Bloodstream infection occurred in 2 of those 91 FP-treated patients.

Dosing: Schedule A: Every 72 h (Day 0–14), every 48 h (Day 15–28), and every 24 h (Day 29–42).

Schedule B: Twice weekly

Schedule C: Every 72 h (Day 0–7), every q24 h (Day 8–28)

Schedule D: Every 72 h (Day 0–7), then every q48 h.

Schedule E: Every 72 h (Day 0–14), then every q48 h.

Schedule F: Every 48 h (Day 0–14), then every q24 h.

The first RCT targeted high-risk infants weighing less than 1000 g with a central venous catheter or endotracheal tube in place during the first five days of life (mean time of enrollment being less than 48 hours old) and continued prophylaxis while they still needed intravenous access (peripheral or central) for up to 6 weeks (Kaufman et al., 2001). ICI occurred in none of the fluconazole-treated, compared with 20% of the placebo-treated patients ($P = 0.008$). Sub-analysis demonstrated efficacy even in the most extremely preterm infants. Nine percent of study infants were less than 24 weeks gestation and fluconazole prophylaxis was effective in preventing ICI in this group ($P = 0.04$). Furthermore, even when these extremely high-risk patients were removed from the analysis, fluconazole prophylaxis was still effective in decreasing the incidence of ICI ($P = 0.01$). Recently, a multi-center RCT in 322 neonates weighing less than 1500 g, enrolled by 48 h of life, demonstrated a decrease in the ICI in the fluconazole-treated infants (odds ratio 0.25; 95% confidence interval, 0.10–0.59; $P = 0.001$) (Manzoni et al., 2007b). In a secondary analysis, prophylaxis was efficacious for infants of less than 27 weeks gestation ($P = 0.007$) and <1000 g ($P = 0.02$).

An important feature of these RCTs was the use primarily of a different anti-fungal agent (amphotericin B) than fluconazole for treatment of documented fungal infections (as well as suspected or empiric treatment) for any infant in the NICU. This reduces the overall exposure of fungi to fluconazole and the potential of resistance to develop (Clancy et al., 2006). It also assures appropriate antifungal therapy if an infant on prophylaxis develops an infection due to a resistant organism.

10.2 Combined Outcome of ICI or Mortality

Analysis of all the randomized controlled trials demonstrates a significant decrease in the combined outcome of ICI or mortality which occurred in 34 of 319 (10%) fluconazole-treated patients, compared to 51 of 206 (25%) in the placebo group (odds ratio 0.36; 95% confidence interval, 0.23–0.58; $P < 0.0001$) (Kaufman, 2008).

10.3 Meta-Analysis and Cochrane Review

Meta-analysis using Mantel-Haenszel methods of the studies from 2001 to 2007 demonstrated that fluconazole prophylaxis reduced the risk of developing ICI in high-risk infants weighing less than 1000 g by 91% (odds ratio 0.09; 95% confidence interval, 0.04–0.24; $P = 0.0004$) and all infants weighing less than 1500 g by 85% (odds ratio 0.15; 95% confidence interval, 0.08–0.26; $P < 0.0001$) (Kaufman et al., 2001; Manzoni et al., 2006a; Kaufman et al., 2005; Kicklighter et al., 2001; Healy et al., 2005; Bertini et al., 2005; Uko et al., 2006; Aghai et al., 2006; Kaufman et al., 2008c; Manzoni et al., 2007b). *Candida*-related mortality was decreased by 96% (odds ratio 0.04; 95% confidence interval, 0.01–0.31; $P = 0.0055$) and overall mortality rate by 25% (11% in the fluconazole-treated infants compared with 16.3% in the control patients) (odds ratio 0.75; 95% confidence interval, 0.58–0.97; $P = 0.029$). Healy et al. also reported the elimination of *Candida*-related mortality in any patient in their NICU when fluconazole prophylaxis was targeted to infants weighing less than 1000 g (Healy et al., 2005; 2008). Finally, the Cochrane data revealed a statistically significant reduction in ICI in infants who received prophylaxis (typical relative risk: 0.23; 95% confidence interval, 0.11–0.46), with the number needed to treat of nine (95% confidence interval 6–17) (Clerihew et al., 2007).

10.4 Resistance and Safety

Some of the issues related to prophylaxis include side effects and resistance. With any antimicrobial therapy, resistance needs to be examined. There are important differences between development of resistance in bacteria and fungi. Most importantly, fungi do not have mechanisms of transferring resistance to other fungi and lower dosing is not associated with emergence of resistance (White et al., 1998; Hof, 2008). In bone marrow transplant patients fluconazole prophylaxis has been successful for nearly 20 years in decreasing both ICI and mortality, while azole-resistant isolates have remained low (approximately 5%) (Marr et al., 2000a; Marr et al., 2000b). Resistance has been observed in adult studies (HIV-positive patients) with prolonged courses (3–24 months) of fluconazole treatment and when escalating to higher doses (400–800 mg) (White et al., 1998). In contrast, adult studies employing lower doses (150–200 mg) and weekly doses have been successful in preventing fungal infections without the development of resistance (Pappas et al., 2004; Sobel et al., 2004).

Neonatal prophylaxis studies have not reported any significant change or emergence of resistant species over the course of prophylaxis, during the study periods of 2–3 years, or over a five-year period encompassing two studies (Kaufman et al., 2001; 2005). Furthermore, there was no emergence or increase in the incidence of colonization or infection due to *C. glabrata* or *C. krusei* reported in any studies as well as a recent single center ten-year analysis (four years prior to and six years following fluconazole prophylaxis) (Manzoni et al., 2008).

Fluconazole prophylaxis at doses 6 mg/kg or greater and increased frequency may be associated with the development of some resistance (Sarvikivi et al., 2005; Yoder et al., 2004). Two studies in preterm infants and one in preterm baboons have reported this association (Parikh et al., 2007; Sarvikivi et al., 2005; Yoder et al., 2004). Sarvikivi et al. have reported on the use of fluconazole prophylaxis over a 12-year period, finding no *C. glabrata* or *C. krusei* infections and only two cases of resistant *C. parapsilosis* BSIs. In their NICU, between 1991 and 2000, fluconazole prophylaxis was most commonly administered daily in doses of 6–12 mg/kg to infants weighing less than 1000 g and less than 30 weeks gestation. No fungal resistance was detected during that time period, however, in 2000, when the use of fluconazole was broadened to the entire NICU, the emergence of resistance was noted and associated with the two BSIs (Sarvikivi et al., 2005). Both patients were infected with the endemic *C. parapsilosis* strain that had previously been suscepti-ble. One important distinction in this series was that fluconazole was used both for prophylaxis and treatment of fungal infections in this NICU. Furthermore, higher doses of daily fluconazole are associated with development of resistance and may have been a factor in this study (White et al., 1998). These observations suggest that lower and less frequent dosing (3 mg/kg twice a week) may delay or attenuate the development of resistance and still prevent serious infections (Kaufman et al., 2005).

Parikh et al. recently reported in a single center study that fluconazole prophy-laxis (at 6 mg/kg every 72 h until day seven, then daily until day 28) decreased colonization but not ICI, with non-albicans species responsible for 96.8% of infec-tions and the majority due to *C. glabrata*. Prior to their prophylaxis study period, fluconazole was used for treatment in their NICU with rates of ICI at approx-imately 25% in infants weighing less than 1500 g, meaning that there was a high exposure rate of treatment doses per patient and it is possible that this total dose exposure in their NICU correlated with the selection of *C. glabrata* or other species with decreased susceptibility to fluconazole. Therefore, if a NICU has been primarily using fluconazole for treatment or empiric therapy and has high rates of infection, fluconazole prophylaxis may not be effective. Changing primary treatment and empiric therapy to one of the amphotericin B formu-lations for 6–12 months before instituting prophylaxis may change the NICU *Candida* flora, as resistance to azoles decreases when fungi are not continuously exposed to them (Hof, 2008). Additionally, amphotericin B may also eradicate some of the resistant species from the horizontally transmitted flora. In those NICUs, nystatin prophylaxis may be a better alternative, and its efficacy further studied.

These findings suggest that it is important to focus fluconazole prophylaxis use in only select high-risk NICU patients and primarily for prophylaxis, with a different antifungal chosen for treatment or empiric therapy, such as amphotericin B when able, as this will limit overall fungal exposure to fluconazole, and possibly prevent the emergence of resistance (Clancy et al., 2006).

10.5 Side Effects of Fluconazole Prophylaxis

In the randomized controlled trials (RCTs) there were no significant differences in bacterial infections, necrotizing enterocolitis, or liver function. There are no adverse effects on long-term neurodevelopmental outcomes (Kaufman et al., 2003). One of the retrospective studies reported a higher incidence of cholestasis in the flu-conazole prophylaxis patients that was transient, with no difference at discharge, while another retrospective study demonstrated a lower incidence of cholestasis

(Uko et al., 2006; Aghai et al., 2006). Since there were no significant differences in direct bilirubin or liver enzymes in the four randomized placebo-controlled trials, it may be that other factors present during the study period increased the likelihood of cholestasis (Healy et al., 2008).

10.6 Cost-Benefit Ratio

Fluconazole prophylaxis is extremely cost effective. Uko et al. examined the cost with fluconazole prophylaxis and showed a significant cost benefit of U.S. $516,702 over 18 months in their NICU (Uko et al., 2006). At our institution, pharmacy cost of one dose is $18 (Marcia Buck, PharmD, personal communication), making the cost of the average time of prophylaxis of 4–6 weeks (8–12 doses) between $144 and $216, per patient.

11 Nystatin

Non-absorbable nystatin prophylaxis is used to prevent or reduce gastrointestinal tract fungal colonization. The first RCT of nystatin prophylaxis (100,000 units orally, three times a day) was studied in ventilated preterm infants weighing less than 1250 g until 1 week after extubation and demonstrated a decrease in *Candida* UTI, but was underpowered to show an effect on the incidence of fungal BSI on subanalysis (Sims et al., 1988). *Candida* BSI occurred in none of the 33 nystatin-treated patients compared to two of 34 (6%) placebo-treated patients ($P = 0.25$). Fungal UTI occurred in two (6%) nystatin and 10 (29%) control patients ($P = 0.01$).

A recent prospective single-center study examined nystatin prophylaxis (100,000 units three times a day orally, or via nasogastric tube) in both preterm and term NICU patients, excluding infants with congenital defects requiring surgery (Ozturk et al., 2006). Examining the study infants by birth weight, 24% ($n = 948$) were less than 1500 g; 8.7% ($n = 349$) were less than 1000 g and 0.8% ($n = 30$) were less than 750 g. Infants were randomized into two groups: Group A had oral fungal cultures performed and received nystatin prophylaxis only if oral *Candida* colonization was isolated. In Group B, all infants were given prophylaxis, regardless of colonization status. In Group A, infants weighing less than 1500 g who did not receive nystatin prophylaxis had a 36% rate of candidemia (131 of 358), compared with an incidence of 13.9% (16 of 115) of Group A infants given prophylaxis only if oral colonization was detected ($p \leq 0.001$). In Group B, where nystatin prophylaxis was started in the first 72 h of life, regardless of the colonization status, only 3.6% (17 of 475) developed invasive infection ($p \leq 0.001$ compared to the Group A infants who received nystatin prophylaxis if colonization was detected). Nystatin was started at 12.6 ± 2.4 (mean $\pm$ sd) days in the subgroup of Group A receiving prophylaxis, compared to 2 ± 0.5 days in Group B. Prophylaxis was shown not to decrease infection in infants weighing more than 1500 g at birth. Nystatin prophylaxis initiated in the first 72 h of life showed lower rates of infection, compared to starting nystatin prophylaxis after oral colonization was detected (3.6% vs. 13.9%; $p = 0.01$), demonstrating the added benefit of using prophylaxis early after birth and thus prior to most colonization. This decreased efficacy of nystatin prophylaxis when initiated after colonization has also been reported in other studies (Leibovitz et al., 1992; Weitkamp et al., 1998; El Masry et al., 2002; Johnson et al., 1984). Limitations of the study included the lack of a placebo group. There were few infants weighing less than 750 g, and infants with congenital defects (e.g. gastroschisis) requiring surgery were excluded. These are high-risk groups in which efficacy needs to be studied.

12 Fluconazole or Nystatin?

From the data above, there is overwhelming evidence (Class 1) to support fluconazole prophylaxis and is consistent with the recent Cochrane review (Kaufman, 2008c; Clerihew et al., 2007). Further study of nystatin prophylaxis is needed in extremely preterm infants, as we only have one RCT in intubated infants weighing less than 1250 g.

12.1 Class 1 Evidence for the Use of Fluconazole Prophylaxis

The studies and meta-analysis of the fluconazole prophylaxis would be classified as Class 1 Evidence defined as (1) definitely recommended; (2) definitive excellent evidence provides support from one or more Level 1 Studies (RCTs or meta-analyses of multiple clinical trials with substantial treatment effects) with (2) Study results consistently positive and compelling, always acceptable and safe, and definitely useful. From the evidence relating safety and resistance, it is important to focus fluconazole prophylaxis in selected high-risk NICU patients, during a high-risk period for infection, thereby limiting patient and fungal exposure.

13 Deciding Who is High-Risk

It is important to examine one's own infection rates in each NICU (Table 3). Epidemiologic studies have lacked critical information regarding antifungal prophylaxis, and data by each gestational age by week and 100-gram birth weight group needs to be analyzed. Without this information, these studies may have minimized the incidence of ICI if infants were receiving antifungal prophylaxis, and may also have missed important risk factors. Nystatin prophylaxis in preterm infants is common and may explain the lower reported rates in the UK and Australia, as well as the varying incidence among United States NICUs (Clerihew and McGuire, 2007; Burwell et al., 2006; O'Grady and Dempsey, 2008).

14 Who Should Receive Antifungal Prophylaxis?

The question many have raised is, *who would benefit from receiving antifungal prophylaxis?* (Burwell et al., 2006) (Table 6). Approximately 30–34% of NICUs have instituted antifungal prophylaxis in the United States and the UK, while greater than 50% use it in Ireland (Clerihew and McGuire, 2007; Burwell et al., 2006; O'Grady and Dempsey, 2008). Several factors should go into this decision; among them, incidence, mortality and neurodevelopmental impairment.

Table 6 Fluconazole prophylaxis algorithm for preterm infants

High risk groups	<1000 g birth weight OR ≤27 weeks gestation	1000–1500 g birth weight
Criteria	• <5 days of life • Endotracheal tube OR • Central venous catheter	• Antibiotic therapy for >3 days • With Central venous catheters
Dosing	3 mg/kg IV Fluconazole Twice a week	
Length of prophylaxis	Up to Day of life 42 Prophylaxis will be stopped prior to 6 weeks if: • No need for IV (peripheral or central) access • Initiation of treatment of documented invasive fungal infection	• During antibiotic treatment • While Central venous catheter is in place
Monitoring	• Weekly or every other week liver function testing[a] • Susceptibility testing of all clinical isolates[b]	
Level of evidence	Based on randomized placebo-controlled trials A-I	Retrospective studies Needs further study B-II

Treatment of documented or suspected invasive fungal infections with non-azole: amphotericin B deoxycholate (1 mg/kg daily) or amphotericin B lipid formulations (5 mg/kg daily).

[a] Liver function testing: direct bilirubin, g-glutamyltransferase, aspartate aminotransferase, alanine aminotransferase; alkaline phosphatase.

[b] Surveillance cultures at initiation and upon completion optional: Less than or equal to 7 days of life: skin, rectal or stool, umbilicus. More than 7 days of life: skin, rectal or stool, endotracheal tube if intubated, nasopharynx if not intubated.

(i) *Targeted prophylaxis should be given to all infants weighing less than 1000 g and/or 27 weeks or younger, while they require IV access (peripheral or central), starting on day of life one and up to 6 weeks of life. (A-I Evidence)*

This subpopulation of preterm infants has high mortality and neurodevelopmental impairment, and this approach has demonstrated efficacy and safety with fluconazole, without the emergence of resistance in randomized placebo controlled trials, while eliminating *Candida*-related mortality.

(ii) *Even in a NICU with overall low rates of ICI (<2%), infants 26 weeks or younger are likely high-risk and would benefit from prophylaxis. Incidence and outcomes by gestational age should be examined and tracked* (Table 3).

ICI can be analyzed by filling out Table 3 at institutions with low rates to determine the gestational age range in which ICI does occur and to identify those infants who should receive prophylaxis. There is likely a gestational age cutoff wherein ICI falls to zero. If NICUs do not have neurodevelopmental outcome data, prophylactic treatment of high risk infants weighing less than 1000 g or 27 weeks *or younger* should be strongly be considered, as treatment of documented infections does not always prevent the neurodevelopmental impairment and mortality of these infections (Stoll et al., 2004; Benjamin et al., 2006).

(iii) *NICUs with high rates in infants weighing between 1000 and 1500 g, may choose prophylaxis in these infants. (B-II Evidence)*

A targeted approach to infants with a central venous catheter or on antibiotics for more than 3 days has been used in retrospective studies (Bertini et al., 2005; Uko et al., 2006).

15 Dosage and Schedule

Fluconazole administered at 3 mg/kg, intravenously, twice a week until IV access (peripheral or central) is no longer needed has the most efficacy and best safety profile. Manzoni et al. in their multi-center RCT, demonstrated that a dosage of 3 or 6 mg/kg is equally effective (Manzoni et al., 2007). However, dosing with 3 mg/kg is preferable for two reasons: 1. drug concentrations in the skin, lung and mucous membranes are greater than plasma levels (therefore larger doses may be unnecessary), and 2. the use of higher doses may foster the development of fungal resistance; furthermore, the goal of prophylaxis to use the lowest effective dose (usually 50% of the treatment dose). At our institution, we administer fluconazole prophylaxis on admission and then twice weekly, every Tuesday and Friday at a specified time (e.g. 10:00 AM), which further reduces pharmacy costs and limits medication errors. For example, a 26-week old infant who reaches full feedings and has his central venous catheter removed at 3 weeks of age, would only need to receive six doses of fluconazole. Therefore, 3 mg/kg given twice a week is the optimal dosing schedule, maximizing efficacy, safety and cost.

16 Summary

Pediatrics has led the way in infectious-disease prevention and now this can benefit one cause of nosocomial infection in preterm infants and it should be instituted in every NICU. Pediatric, infectious diseases, and neonatal organizations can help guide NICUs regarding antifungal prophylaxis, and support good epidemiologic follow-up of ICI, mortality, and fungal susceptibilities. The 2006 Report of the *Committee on Infectious Diseases of the American Academy of Pediatrics* (Red Book®) features a statement supporting the use of fluconazole prophylaxis in high-risk preterm infants weighing less than 1000 g (*Candidiasis in Eds.* Pickering et al., 2006).

In striving for better outcomes for our most extremely preterm infants, from the delivery room to discharge, infection prevention is critical to decreasing many of the associated morbidities and mortality. At this time, the benefits of antifungal prophylaxis significantly outweigh any risks. With single- and multi-center randomized controlled studies and a meta-analysis demonstrating a 91% decrease of ICI in infants weighing less than 1000 g and 96% decrease in infection-related mortality, fluconazole prophylaxis should be targeted at this group of infants weighing less than 1000 g or 27 weeks or younger, due to the high mortality rate and neurodevelopmental impairment. The prevention of ICI in extremely preterm infants also eliminates *Candida* as a cause of mortality, as well as neurodevelopmental impairment in these vulnerable hosts.

References

Aghai, Z.H., Mudduluru, M., Nakhla, T.A., Amendolia, B., Longo, D., Kemble, N., Kaki, S., Sutsko, R., Saslow, J.G., & Stahl, G.E. (2006). Fluconazole prophylaxis in extremely low birth weight infants: association with cholestasis. *J Perinatol,* (26), 550–555.

Benjamin, D.K., DeLong, E., Cotten, C.M., Garges, H.P., Steinbach, W.J., & Clark, R.H. (2004). Mortality following blood culture in premature infants: increased with gram-negative bacteremia and candidemia, but not gram-positive bacteremia. *J Perinatol,* (24), 175–180.

Benjamin, D.K., Jr., DeLong, E.R., Steinbach, W.J., Cotton, C.M., Walsh, T.J., & Clark, R.H. (2003). Empirical therapy for neonatal candidemia in very low birth weight infants. *Pediatrics,* (112), 543–547.

Benjamin, D.K., Jr., Stoll, B.J., Fanaroff, A.A., McDonald, S.A., Oh, W., Higgins, R.D., Duara, S., Poole, K., Laptook, A., & Goldberg, R. (2006). Neonatal candidiasis among extremely low birth weight infants: risk factors, mortality rates, and neurodevelopmental outcomes at 18 to 22 months. *Pediatrics,* 117(1), 84–92.

Bertini, G., Perugi, S., Dani, C., Filippi, L., Pratesi, S., & Rubaltelli, F.F. (2005). Fluconazole prophylaxis prevents invasive fungal infection in high-risk, very low birth weight infants. *J Pediatr,* (147), 162–165.

Burwell, L.A., Kaufman, D., Blakely, J., Stoll, B.J., & Fridkin, S.K. (2006). Antifungal prophylaxis to prevent neonatal candidiasis: a survey of perinatal physician practices. *Pediatrics,* (118), e1019–e1026.

Clancy, C.J., Staley, B., & Nguyen, M.H. (2006). In vitro susceptibility of breakthrough *Candida* bloodstream isolates correlates with daily and cumulative doses of fluconazole. *Antimicrob Agents Chemother,* (50), 3496–3498.

Clerihew, L., Austin, N., & McGuire, W. (2007). Prophylactic systemic antifungal agents to prevent mortality and morbidity in very low birth weight infants. *Cochrane Database Syst Rev,* CD003850.

Clerihew, L., Lamagni, T.L., Brocklehurst, P., & McGuire, W. (2006). Invasive fungal infection in very low birthweight infants: national prospective surveillance study. *Arch Dis Child Fetal Neonatal Ed,* (91), F188–F192.

Clerihew, L. & McGuire, W. (2007). Antifungal prophylaxis for very-low-birth-weight infants: UK national survey. *Arch Dis Child Fetal Neonatal Ed,* 93(3), F238–F239.

Cotten, C.M., McDonald, S., Stoll, B., Goldberg, R.N., Poole, K., & BenjaminJr., D.K. (2006). The association of third-generation cephalosporin use and invasive candidiasis in extremely low birth-weight infants. *Pediatrics,* (118), 717–722.

El Masry, F.A., Neal, T.J., & Subhedar, N.V. (2002). Risk factors for invasive fungal infection in neonates. *Acta Paediatr,* (91), 198–202.

Ellepola, A.N. & Samaranayake, L.P. (1998). Adhesion of oral *C. albicans* to human buccal epithelial cells following limited exposure to antifungal agents. *J Oral Pathol Med,* (27), 325–332.

Faergemann, J. & Laufen, H. (1993). Levels of fluconazole in serum, stratum corneum, epidermis-dermis (without stratum corneum) and eccrine sweat. *Clin Exp Dermatol,* (18), 102–106.

Fairchild, K.D., Tomkoria, S., Sharp, E.C., & Mena, F.V. (2002). Neonatal *Candida glabrata* sepsis: clinical and laboratory features compared with other *Candida* species. *Pediatr Infect Dis J,* (21), 39–43.

Feja, K.N., Wu, F., Roberts, K., Loughrey, M., Nesin, M., Larson, E., la-Latta, P., Haas, J., Cimiotti, J., & Saiman, L. (2005). Risk factors for candidemia in critically Ill infants: a matched case-control study. *J Pediatr,* (147), 156–161.

Fridkin, S.K., Kaufman, D., Edwards, J.R., Shetty, S., & Horan, T. (2006). Changing incidence of *Candida* bloodstream infections among NICU patients in the United States: 1995–2004. *Pediatrics,* (117), 1680–1687.

Ghannoum, M.A., Filler, S.G., Ibrahim, A.S., Fu, Y., & Edwards, J.E., Jr. (1992). Modulation of interactions of *Candida albicans* and endothelial cells by fluconazole and amphotericin B. *Antimicrob Agents Chemother,* (36), 2239–2244.

Guillet, R., Stoll, B.J., Cotten, C.M., Gantz, M., McDonald, S., Poole, W.K., & Phelps, D.L. (2006). Association of H2-blocker therapy and higher incidence of necrotizing enterocolitis in very low birth weight infants. *Pediatrics,* (117), e137–e142.

Hack, M. (2006). Neonatology fellowship training in research pertaining to development and follow-up. *J Perinatol,* 26(Suppl 2), S30–S33.

Hazen, K.C., Coleman, E., & Wu, G. (2000). Influence of fluconazole at subinhibitory concentrations on cell surface hydrophobicity and phagocytosis of *Candida albicans*. *FEMS Microbiol Lett,* (183), 89–94.

Healy, C.M., Baker, C.J., Zaccaria, E., & Campbell, J.R. (2005). Impact of fluconazole prophylaxis on incidence and outcome of invasive candidiasis in a neonatal intensive care unit. *J Pediatr,* (147), 166–171.

Healy, C.M., Campbell, J.R., Zaccaria, E., & Baker, C.J. (2008). Fluconazole prophylaxis in extremely low birth weight neonates reduces invasive candidiasis mortality rates without emergence of fluconazole-resistant *Candida* species. *Pediatrics,* (121), 703–710.

Hof, H. (2008). Is there a serious risk of resistance development to azoles among fungi due to the widespread use and long-term application of azole antifungals in medicine? *Drug Resist Updat,* 11(1–2), 25–31.

Houang, E.T., Chappatte, O., Byrne, D., Macrae, P.V., & Thorpe, J.E. (1990). Fluconazole levels in plasma and vaginal secretions of patients after a 150-milligram single oral dose and rate of eradication of infection in vaginal candidiasis. *Antimicrob Agents Chemother,* (34), 909–910.

Hylander, M.A., Strobino, D.M., & Dhanireddy, R. (1998). Human milk feedings and infection among very low birth weight infants. *Pediatrics,* (102), E38.

Johnson, D.E., Thompson, T.R., Green, T.P., & Ferrieri, P. (1984). Systemic candidiasis in very low-birth-weight infants (less than 1,500 grams). *Pediatrics,* (73), 138–143.

Johnsson, H. & Ewald, U. (2004). The rate of candidaemia in preterm infants born at a gestational age of 23–28 weeks is inversely correlated to gestational age. *Acta Paediatr,* (93), 954–958.

Kaufman, D.A. (2008a). Prevention of invasive *Candida* infections in preterm infants: the time is now. *Expert Rev Anti Infect Ther,* (6), 393–399.

Kaufman, D. (2008b). Fluconazole prophylaxis decreases the combined outcome of invasive *Candida* infections or mortality in preterm infants. *Pediatrics,* (122), 1158–1159.

Kaufman, D.A. (2008c). Fluconazole prophylaxis: can we eliminate invasive *Candida* infections in the neonatal ICU? *Curr Opin Pediatr,* 20(3), 332–340.

Kaufman, D., Boyle, R., Hazen, K.C., Patrie, J.T., Robinson, M., & Donowitz, L.G. (2001). Fluconazole prophylaxis against fungal colonization and infection in preterm infants. *N Engl J Med,* (345), 1660–1666.

Kaufman, D., Boyle, R., Hazen, K.C., Patrie, J.T., Robinson, M., & Grossman, L.B. (2005). Twice weekly fluconazole prophylaxis for prevention of invasive *Candida* infection in high-risk infants of <1000 grams birth weight. *J Pediatr,* (147), 172–179.

Kaufman, D., Boyle, R., Robinson, M., & Grossman, L.B. (2003). Long-term safety of intravenous prophylactic fluconazole use in preterm infants less than 1000 grams. *Pediatr Res,* (53), 484A. Ref Type: Abstract.

Kaufman, D.A., Gurka, M.J., Hazen, K.C., Boyle, R., Robinson, M., & Grossman, L.B. (2006). Patterns of fungal colonization in preterm infants weighing less than 1000 grams at birth. *Pediatr Infect Dis J,* (25), 733–737.

Kaufman, D.A., Manzoni, P., Gurka, M.J., & Grossman, L.B. (2007). Antifungal Prophylaxis in the Neonatal Intensive Care Unit (NICU). *Curr Pediatr Rev,* 277–288.

Kicklighter, S.D., Springer, S.C., Cox, T., Hulsey, T.C., & Turner, R.B. (2001). Fluconazole for prophylaxis against candidal rectal colonization in the very low birth weight infant. *Pediatrics,* (107), 293–298.

Kojic, E.M. & Darouiche, R.O. (2004). *Candida* infections of medical devices. *Clin Microbiol Rev,* (17), 255–267.

Koks, C.H., Crommentuyn, K.M., Hoetelmans, R.M., Mathot, R.A., & Beijnen, J.H. (2001). Can fluconazole concentrations in saliva be used for therapeutic drug monitoring? *Ther Drug Monit,* (23), 449–453.

Kramer, M.S., Platt, R.W., Wen, S.W., Joseph, K.S., Allen, A., Abrahamowicz, M., Blondel, B., & Breart, G. (2001). A new and improved population-based Canadian reference for birth weight for gestational age. *Pediatrics,* (108), E35.

Leibovitz, E., Iuster-Reicher, A., Amitai, M., & Mogilner, B. (1992). Systemic candidal infections associated with use of peripheral venous catheters in neonates: a 9-year experience. *Clin Infect Dis,* (14), 485–491.

Makhoul, I.R., Bental, Y., Weisbrod, M., Sujov, P., Lusky, A., & Reichman, B. (2007). Candidal versus bacterial late-onset sepsis in very low birthweight infants in Israel: a national survey. *J Hosp Infect,* (65), 237–243.

Makhoul, I.R., Sujov, P., Smolkin, T., Lusky, A., & Reichman, B. (2002). Epidemiological, clinical, and microbiological characteristics of late-onset sepsis among very low birth weight infants in Israel: a national survey. *Pediatrics,* (109), 34–39.

Makhoul, I.R., Sujov, P., Smolkin, T., Lusky, A., & Reichman, B. (2005). Pathogen-specific early mortality in very low birth weight infants with late-onset sepsis: a national survey. *Clin Infect Dis,* (40), 218–224.

Manzoni, P., Arisio, R., Mostert, M., Leonessa, M., Farina, D., Latino, M.A., & Gomirato, G. (2006a). Prophylactic fluconazole is effective in preventing fungal colonization and fungal systemic infections in preterm neonates: a single-center, 6-year, retrospective cohort study. *Pediatrics,* (117), e22–e32.

Manzoni, P., Farina, D., Antonielli, D.E., Leonessa, M.L., Gomirato, G., & Arisio, R. (2006b). An association between anatomic site of *Candida* colonization and risk of invasive candidiasis exists also in preterm neonates in neonatal intensive care unit. *Diagn Microbiol Infect Dis,* (56), 459–460.

Manzoni, P., Farina, D., Galletto, P., Leonessa, M., Priolo, C., Arisio, R., & Gomirato, G. (2007a). Type and number of sites colonized by fungi and risk of progression to invasive fungal infection in preterm neonates in neonatal intensive care unit. *J Perinat Med,* (35), 220–226.

Manzoni, P., Farina, D., Leonessa, M., d'Oulx, E.A., Galletto, P., Mostert, M., Miniero, R., & Gomirato, G. (2006c). Risk factors for progression to invasive fungal infection in preterm neonates with fungal colonization. *Pediatrics,* (118), 2359–2364.

Manzoni, P., Leonessa, M., Galletto, P., Latino, M.A., Arisio, R., Maule, M., Agriesti, G., Gastaldo, L., Gallo, E., Mostert, E., & Farina, D. (2008). Routine use of fluconazole prophylaxis in a neonatal intensive care unit does not select natively fluconazole-resistant *Candida* subspecies. *Pediatr Infect Dis J,* (27), 731–737.

Manzoni, P., Mostert, M., Leonessa, M.L., Priolo, C., Farina, D., Monetti, C., Latino, M.A., & Gomirato., G. (2006d). Oral supplementation with Lactobacillus casei subspecies rhamnosus prevents enteric colonization by *Candida* species in preterm neonates: a randomized study. *Clin Infect Dis,* (42), 1735–1742.

Manzoni, P., Stolfi, I., Pugni, L., Decembrino, L., Magnani, C., Vetrano, G., Tridapalli, E., Corona, G., Giovannozzi, C., Farina, D., Arisio, R., Merletti, F., Maule, M., Mosca, F., Pedicino, R., Stronati, M., Mostert, M., & Gomirato, G. (2007b). A multi-center, randomized trial of prophylactic fluconazole in preterm neonates. *N Engl J Med,* (356), 2483–2495.

Marr, K.A., Seidel, K., Slavin, M.A., Bowden, R.A., Schoch, H.G., Flowers, M.E., Corey, L., & Boeckh, M. (2000b). Prolonged fluconazole prophylaxis is associated with persistent protection against candidiasis-related death in allogeneic marrow transplant recipients: long-term follow-up of a randomized, placebo-controlled trial. *Blood,* (96), 2055–2061.

Marr, K.A., Seidel, K., White, T.C., & Bowden, R.A. (2000a). Candidemia in allogeneic blood and marrow transplant recipients: evolution of risk factors after the adoption of prophylactic fluconazole. *J Infect Dis,* (181), 309–316.

Ozturk, M.A., Gunes, T., Koklu, E., Cetin, N., & Koc, N. (2006). Oral nystatin prophylaxis to prevent invasive candidiasis in Neonatal Intensive Care Unit. *Mycoses,* (49), 484–492.

O'Grady, M.J. & Dempsey, E.M. (2008). Antifungal prophylaxis for the prevention of neonatal candidiasis? *Acta Paediatr,* (97), 430–433.

Pappas, P.G., Rex, J.H., Sobel, J.D., Filler, S.G., Dismukes, W.E., Walsh, T.J., & Edwards, J.E. (2004). Guidelines for treatment of candidiasis. *Clin Infect Dis,* (38), 161–189.

Parikh, T.B., Nanavati, R.N., Patankar, C.V., Suman Rao, P.N., Bisure, K., Udani, R.H., & Mehta, P. (2007). Fluconazole prophylaxis against fungal colonization and invasive fungal infection in very low birth weight infants. *Indian Pediatr,* (44), 830–837.

Candidiasis. In Pickering, L., Baker, C.J., Long, S.S., & McMillian, J.A., (Eds.) (2006). *Red Book. Report of the Committee on Infectious Diseases.* (27th ed.). Elk Grove Villiage, IL: American Academy of Pediatrics. pp. 246.

Procianoy, R.S., Eneas, M.V., & Silveira, R.C. (2006). Empiric guidelines for treatment of *Candida* infection in high-risk neonates. *Eur J Pediatr,* (165), 422–423.

Saiman, L., Ludington, E., Dawson, J.D., Patterson, J.E., Rangel-Frausto, S., Wiblin, R.T., Blumberg, H.M., Pfaller, M., Rinaldi, M., Edwards, J.E., Wenzel, R.P., & Jarvis, W. (2001). Risk factors for *Candida* species colonization of neonatal intensive care unit patients. *Pediatr Infect Dis J,* (20), 1119–1124.

Saiman, L., Ludington, E., Pfaller, M., Rangel-Frausto, S., Wiblin, R.T., Dawson, J., Blumberg, H.M., Patterson, J.E., Rinaldi, M., Edwards, J.E., Wenzel, R.P., & Jarvis, W. (2000). Risk factors for candidemia in neonatal intensive care unit patients. The National Epidemiology of Mycosis Survey study group. *Pediatr Infect Dis J,* (19), 319–324.

Sarvikivi, E., Lyytikainen, O., Soll, D.R., Pujol, C., Pfaller, M.A., Richardson, M., Koukila-Kahkola, P., Luukkainen, P., & Saxen, H. (2005). Emergence of fluconazole resistance in a *Candida* parapsilosis strain that caused infections in a neonatal intensive care unit. *J Clin Microbiol,* (43), 2729–2735.

Schuman, P., Capps, L., Peng, G., Vazquez, J., el Sadr, W., Goldman, A.I., Alston, B., Besch, C.L., Vaughn, A., Thompson, M.A., Cobb, M.N., Kerkering, T., & Sobel, J.D. (1997). Weekly fluconazole for the prevention of mucosal candidiasis in women with HIV infection. A randomized, double-blind, placebo-controlled trial. Terry Beirn Community Programs for Clinical Research on AIDS. *Ann Intern Med,* (126), 689–696.

Sims, M.E., Yoo, Y., You, H., Salminen, C., & Walther, F.J. (1988). Prophylactic oral nystatin and fungal infections in very-low- birthweight infants. *Am J Perinatol,* (5), 33–36.

Smith, P.B., Morgan, J., Benjamin, J.D., Fridkin, S.K., Sanza, L.T., Harrison, L.H., Sofair, A.N., Huie-White, S., & Benjamin, D.K., Jr. (2007). Excess costs of hospital care associated with neonatal candidemia. *Pediatr Infect Dis J,* (26), 197–200.

Sobel, J.D., Wiesenfeld, H.C., Martens, M., Danna, P., Hooton, T.M., Rompalo, A., Sperling, M., Livengood, C., III, Horowitz, B., Von Thron, J., Edwards, L., Panzer, H., & Chu, T.C. (2004). Maintenance fluconazole therapy for recurrent vulvovaginal candidiasis. *N Engl J Med,* (351), 876–883.

Stoll, B.J., Hansen, N.I., Adams-Chapman, I., Fanaroff, A.A., Hintz, S.R., Vohr, B., & Higgins, R.D. (2004). Neurodevelopmental and growth impairment among extremely low-birth-weight infants with neonatal infection. *JAMA,* (292), 2357–2365.

Stoll, B.J., Hansen, N., Fanaroff, A.A., Wright, L.L., Carlo, W.A., Ehrenkranz, R.A., Lemons, J.A., Donovan, E.F., Stark, A.R., Tyson, J.E., Oh, W., Bauer, C.R., Korones, S.B., Shankaran, S., Laptook, A.R., Stevenson, D.K., Papile, L.A., & Poole, W.K. (2002). Late-onset sepsis in very low birth weight neonates: the experience of the NICHD Neonatal Research Network. *Pediatrics,* (110), 285–291.

Stoll, B.J., Temprosa, M., Tyson, J.E., Papile, L.A., Wright, L.L., Bauer, C.R., Donovan, E.F., Korones, S.B., Lemons, J.A., Fanaroff, A.A., Stevenson, D.K., Oh, W., Ehrenkranz, R.A., Shankaran, S., & Verter, J. (1999). Dexamethasone therapy increases infection in very low birth weight infants. *Pediatrics,* (104), e63.

Uko, S., Soghier, L.M., Vega, M., Marsh, J., Reinersman, G.T., Herring, L., Dave, V.A., Nafday, S., & Brion, L.P. (2006). Targeted short-term fluconazole prophylaxis among very low birth weight and extremely low birth weight infants. *Pediatrics,* (117), 1243–1252.

Vaden, S.L., Heit, M.C., Hawkins, E.C., Manaugh, C., & Riviere, J.E. (1997). Fluconazole in cats: pharmacokinetics following intravenous and oral administration and penetration into cerebrospinal fluid, aqueous humour and pulmonary epithelial lining fluid. *J Vet Pharmacol Ther,* (20), 181–186.

Walsh, T.J., Gonzalez, C.E., Piscitelli, S., Bacher, J.D., Peter, J., Torres, R., Shetti, D., Katsov, V., Kligys, K., & Lyman, C.A. (2000). Correlation between in vitro and in vivo antifungal activities in experimental fluconazole-resistant oropharyngeal and esophageal candidiasis. *J Clin Microbiol,* (38), 2369–2373.

Weitkamp, J.H., Ozdas, A., Lafleur, B., & Potts, A.L. (2008). Fluconazole prophylaxis for prevention of invasive fungal infections in targeted highest risk preterm infants limits drug exposure. *J Perinatol,* (286), 405–411.

Weitkamp, J.H., Poets, C.F., Sievers, R., Musswessels, E., Groneck, P., Thomas, P., & Bartmann, P. (1998). *Candida* infection in very low birth-weight infants: outcome and nephrotoxicity of treatment with liposomal amphotericin B (AmBisome). *Infection,* (26), 11–15.

White, T.C., Marr, K.A., & Bowden, R.A. (1998). Clinical, cellular, and molecular factors that contribute to antifungal drug resistance. *Clin Microbiol Rev,* (11), 382–402.

Yoder, B.A., Sutton, D.A., Winter, V., & Coalson, J.J. (2004). Resistant *Candida* parapsilosis associated with long term fluconazole prophylaxis in an animal model. *Pediatr Infect Dis J,* (23), 687–688.

Zaoutis, T.E., Heydon, K., Localio, R., Walsh, T.J., & Feudtner, C. (2007). Outcomes attributable to neonatal candidiasis. *Clin Infect Dis,* (44), 1187–1193.

Current Status of Treatment of Hepatitis B in Children

Deirdre Kelly

1 Introduction

Viral Hepatitis B is the cause of significant disease worldwide. Acute infection occurs with hepatitis B, but chronic asymptomatic infection leading to chronic liver disease and hepatocellular carcinoma is a life long concern.

The natural history of hepatitis B (HBV) is well established in adults, but the long-term outcome for children is unclear. The main mechanism of infection in childhood is perinatal transmission, which can be prevented effectively by vaccination.

Because of immunotolerance in the host, reduced cell mediated immune response, and the development of viral resistance to oral nucleotide agents, treatment for hepatitis B is rarely effective with current medications. Treatment options for children include the following: interferon (5 mega units per m^2 subcutaneously every 3 weeks for 6 months), lamivudine (3 mg/kg for 12–24 months), or adefovir (10 mg). Interferon clears viral infection in 20–40% of children and is most effective in children with elevated transaminases or horizontal transmission. Only 23% of children seroconvert after lamivudine, 26% of whom many develop resistance with YMDD mutant variants of HBV. Seroconversion rates are only 23% with adefovir in children aged less than 12 years, but viral resistance is not an issue. A number of other drugs, such as entecavair, telbivudine, tenofovir, and pegylated interferon are under evaluation. Children should only be treated as part of a clinical trial or for compassionate reasons.

In children worldwide, infection with viral hepatitis B (HBV) leads to significant disease. Approximately 2 billion people in the world have been infected by HBV, and more than 350 million are chronic carriers. Acute infection is rare in childhood; however, chronic asymptomatic infection, which carries a risk of chronic liver disease and hepatocellular carcinoma is a major concern (Lavanchy, 2004; Chu, 2000).

D. Kelly (✉)
The Liver Unit, Birmingham Children's Hospital, Birmingham, B4 6NH, UK
e-mail: deirdre.kelly@bch.nhs.uk

A. Finn et al. (eds.), *Hot Topics in Infection and Immunity in Children VI*, Advances in Experimental Medicine and Biology 659, DOI 10.1007/978-1-4419-0981-7_10, © Springer Science+Business Media, LLC 2010

In endemic areas, HBV infection takes place in infancy and early childhood with perinatal transmission accounting for most chronic HBV infection. In contrast to infection in adults, HBV infection during early childhood is more likely to lead to persistent infection and a life- time risk of cirrhosis and hepatocellular carcinoma.

Three phases of chronic HBV have been identified: the immune-tolerant phase, the immune-active phase, and the inactive HBV phase. Most children with chronic HBV infection are immune-tolerant, with high viral replication, positive HBV envelope antigen (HBeAg), high HBV deoxyribonucleic acid (DNA) levels, and normal levels of hepatic transaminases.

2 Diagnosis and Natural History

The diagnosis of HBV is made by detecting HBV surface antigen in blood. Acute infection is detected by IgM anticore antibodies to HBV while chronic hepatitis is demonstrated by the presence of IgG anticore HBV and HBV e antigen (HBeAg). Quantitative assay of HBV DNA indicates the level of viral load and determines infectivity.

Following an acute infection, 90% of children will recover spontaneously, while acute fulminant hepatitis develops in less than 1% of patients, some of whom will require liver transplantation (Lee et al., 2005).

The best approach to HBV control is prevention by vaccination and by screening of blood products and organ donors. Recombinant HBV vaccine is effective in 97% of at risk infants. Passive immunisation at birth with HBV immunoglobulin is also required if the mother is HBeAg positive and vaccine is given in all cases at birth with 2 or 3 subsequent vaccinations over 6 months so that all exposed children should receive at least three doses. It is now clear that immunological protection lasts for at least 10 years so that a booster dose may be required for at risk populations (Boxall et al., 2004a, Petersen et al., 2004).

The natural history of chronic HBV infection in childhood varies with the route of infection. The rate of seroconversion is lower in perinatally infected infants (Boxall et al., 2004b) than in horizontally infected infants (Bortolotti et al., 1998). Children are asymptomatic without evidence of chronic liver disease. Biochemical parameters indicate mild elevation of hepatic transaminases (80–150 U/L) with normal albumin, coagulation, and alkaline phosphatase. Liver histology indicates a chronic hepatitis in over 90% of the carriers, which may be mild or nonspecific in 40% of children. There is little correlation between transaminase elevation and the extent of hepatitis (Boxall et al., 2004). Progression to cirrhosis and to hepatocellular carcinoma in childhood has been documented (Moore et al., 2004).

3 Management of Chronic HBV Infection

Annual monitoring of children with persistent HBV should be done clinically and by HBV serology, HBV DNA, standard liver function tests, alpha-fetoprotein, and abdominal ultrasound to detect evidence of seroconversion, progressive liver disease and/or hepatocellular carcinoma. It is essential that the children are encouraged to lead a normal life and are not stigmatised because of their disease. This requires sensitivity from schools and nurseries. Consideration may be given to antiviral therapy.

4 Antiviral Therapy

HBV is not easy to treat, and success rates for viral clearance are approximately 30% in both adults and children. The main reasons for these success rates are the immune tolerant state in which there is an insufficient innate or adaptive immune response to clear the virus (Rehermann, 2003) and the development of viral resistance to therapy. Chronic HBV in children is associated with an HBV-specific T cell hyporesponsiveness and an inadequate CD8 response (Lok and McMahon, 2007). A partial immune response leads to hepatocyte injury and progression of liver disease and is detectable by elevated hepatic transaminases and fibrosis on liver biopsy. It also increases the chance of responding to treatment as discussed below.

The development of viral resistance is related to the combination of host, drug, and viral characteristics. HBV is a small DNA virus in which replication takes place through an RNA intermediate in the hepatocytes within a nucleocapsid that contains the core protein, the pre genomic RNA, and a polymerase enzyme. The HBV polymerase is the main target for anti HBV drugs such as nucleotide and nucleoside analogues. Newer drugs are being developed which will target other steps in viral replication such as the encapsidation step (Ghany and Liang, 2007).

Viral resistance arises because HBV replicates very rapidly and because the HBV reverse transcriptase does not have the function to repair incorrectly incorporated nucleotides, leading to the persistence of viral mutations. The rate of development of mutations is related to how effectively the drug reduces HBV DNA. Thus drug therapy that only partially reduces HBV DNA is more likely to permit mutations as it has been seen with monotherapy with lamivudine (see below). Ideal therapy would act on a range of molecular targets (Ghany and Liang, 2007).

5 Aim of Treatment

Short term treatment aims include the eradication of replicative infection by clearance of HB e antigen, reduction in HBV DNA levels, and normalisation of hepatic transaminases (aspartate transaminases (AST) and alanine trasaminase (ALT)).

Long term aims are focused on preventing the progress of liver disease, on reducing the risk of morbidity and mortality from cirrhosis and hepatocellular carcinoma, and on reducing the pool of carriers. In children and young people, it is important to take into consideration the effect of HBV on their career and marriage prospects and the potential psychosocial stigma.

The decision to treat and the choice of treatment require consideration of the following:

- disease activity: HBV e Ag pos and elevated HBV DNA (greater than 10^5)
- histological stage: hepatitis, not cirrhosis
- likelihood of response (increased transaminases (AST, ALT) more than 2 times normal)
- tolerance of treatment associated side effects
- previous treatment (e.g., resistance to previous therapy)
- co-existing disease (e.g., HIV positivity)

Disease activity before and during treatment is assessed by markers of viral activity such as HBV DNA levels, loss of detectable HB eAg, and markers of immune mediated inflammation such as ALT and AST and histology.

In general, response to treatment is higher in those children with active liver disease/immune response. Nonresponders tend to have low/normal transaminases, persistent viremia on treatment, and insufficient reduction of HBV DNA, which increases the probability of emergence of viral resistance.

6 Indications for Treatment

Consensus guidelines for the treatment of chronic HBV in children have not yet been established and guidelines for antiviral therapy in adults are probably not applicable for children (European Association for the Study, 2009).

The indications for treatment are persistent infection with HBV e Ag positivity for greater than 6 months, evidence of hepatic inflammation either by elevated aminotransferase enzymes or by liver biopsy. Current treatment options include interferon, lamivudine, and adefovir dipivoxil, although a number of other drugs are undergoing evaluation.

6.1 Interferon

Interferon alpha-2b is a naturally occurring protein, which stimulates the immune response by encouraging lymphocyte proliferation, increasing major histocompatability complex antigen expression, and increasing natural killer cell activity. It degrades viral mRNA and inhibits viral protein synthesis. The European consensus on interferon therapy suggests pre and post therapy liver biopsy and interferon 5 megaunits per m^2 subcutaneously 3 times weekly for 6 months (Jara and Bortolotti, 1999). The efficacy of interferon treatment ranges from 20 to 40% and is highest in children with elevated transaminases and those who have been infected by horizontal transmission (Bortolotti et al., 2000). The role of prednisolone priming is unproven, but it is possible that prednisolone increases the spontaneous remission rate and reduces time to seroconversion (Boxall et al., 2006).

Although children tolerate treatment better than adults, interferon has a number of unpleasant side effects. Fever and flu like symptoms are common when treatment is started, as is bone marrow suppression. Autoimmune thyroid disease, alopecia, and mental disturbance, including severe depression may also occur. It is not recommended in children with decompensated liver disease, pancytopenia, and severe renal, cardiac, or autoimmune disease. Pegylated interferon has not yet been approved for HBV treatment in children but is more effective than interferon in adults The covalent attachment of a polyethylene glycol (PEG) moiety enhances the half-life and reduces immunogenicity so that injections can be given once-weekly rather than three times per week (Lau et al., 2005).

6.2 Lamivudine

Lamivudine (also known as 3TC) is a pyrimidine nucleoside analogue that prevents replication of HBV in infected hepatocytes. It is incorporated into viral DNA leading to chain termination and competitively inhibits viral reverse transcriptase. In most patients within 2 weeks of commencing treatment, it leads to a rapid reduction in plasma HBV DNA. Response rates are related to low HBV DNA levels pre-treatment and evidence of hepatic inflammation (raised ALT/AST). A double blind placebo controlled trial of 286 children with chronic HBV, who were treated with 3 mg/kg/day for 12–36 months, showed a complete response, that is, HBe antigen clearance or undetectable HBV DNA after 52 weeks in 23% of children as compared to 13% in the placebo group (Jonas et al., 2002). Treatment for up to 3 years increased the seroconversion rate to 62% in children who did not develop viral mutations, but the high rate of development of YMDD mutant variants of HBV in the majority precludes long-term treatment with lamivudine (Sokal et al., 2006).

6.3 Adefovir Dipivoxil

Adefovir dipivoxil is a purine analogue, which inhibits viral replication by binding to DNA polymerase. It may also augment natural killer cell activity and endogenous interferon activity. Furthermore, HBV strains resistant to lamivudine are susceptible to adefovir (Perillo et al., 1999).

A recent randomized controlled trial in 173 children has demonstrated that adefovir, in doses up to 10 mg, reduced HBV DNA most effectively in children over 12 years of age (23%, which is similar to adults) as compared to younger children. Seroconversion rates were 20% in children aged less than 12 years, but there was no significant difference compared to placebo, although this may have been due to the small numbers of children in the younger cohorts (Jonas et al., 2008). Viral resistance did not occur in the study but is reported in adults in 1% at 1 year, rising to 29% at 5 years and is usually treated using combination therapy with lamivudine. However, there is increased resistance to adefovir in patients who were previously resistant to lamivudine (Hadziyannis et al., 2005). Therefore, except for compassionate use, adefovir is not considered to be first line therapy in young children.

6.4 Combination Therapy

In adults, the combination of pegylated interferon (180 μg weekly) and lamivudine (100 mg daily) for 48 weeks produced greater viral suppression, but no real difference in seroconversion rates, which is disappointing (Lau et al., 2005). There are a few small studies to date of pegylated interferon in children with HBV with or without lamivudine, but the results of these studies are inconclusive (D'Antiga et al., 2006).

6.5 Liver Transplantation

Liver transplantation is effective treatment for children with acute or chronic liver failure. The recurrence of HBV is unusual following transplantation for acute fulminant hepatitis but is usual following transplantation for chronic HBV unless prevented with a combination of oral lamivudine and HBV immune globulin (Terrault, 2002).

7 Future Therapy for HBV in Children

Telbivudine is an L-nucleoside analogue which is more effective than lamivudine (in adults: 26% compared to 23%) but has a high rate of viral resistance compared to adefovir and is not recommended as monotherapy (Lai et al., 2005). A pharmacokinetic study in children aged 2–18 years is being planned.

Tenofovir disoproxil fumarate is a nucleotide analogue similar to adefovir that was originally licensed for treatment of HIV. In vitro studies demonstrated activity against HBV and clinical studies suggest increased potency compared to adefovir (van Bommel et al., 2006). A randomised, placebo controlled multi centre study has recently begun recruitment.

Entecavir, a carbocyclic analogue, inhibits HBV replication at 3 different steps: the priming of HBV DNA polymerase, reverse transcription, and synthesis of HBV-DNA. It is more potent than lamivudine in suppressing wild type HBV, but less effective in adults with lamivudine resistance. Viral resistance is rare. An open label multicentre trial is in progress in children (Sherman et al., 2006).

It is encouraging that EMEA (the European Medicines Agency) is considering establishing guidelines for therapy for viral hepatitis in children and will be planning future therapy.

8 Choice of Therapy for HBV

The management of chronic HBV in asymptomatic children remains a challenge. It is essential that children with HBV disease should be referred to specialised centres to benefit from counselling, to gather information, and for inclusion into multicentre trials of antiviral therapy so that the natural history and outcome of the treatment be appropriately monitored. At present, there is no completely effective therapy, and so all children should be treated within the context of a clinical trial. In children who require compassionate therapy, treatment with lamivudine, which is licensed in children, with or without pegylated interferon is a possibility. If viral mutations develop, then adefovir could be considered until more results are available from current drug trials.

References

Bortolotti, F., Jara, P., Barbera, C., Gregorio, G.V., Veggente, A., Zancan, L., Hierro, L. et al. (2000). Long term effect of alpha interferon in children with chronic hepatitis B. *Gut,* 46(5), 715–718.

Bortolotti, F., Jara, P., Crivello, C., Hierro, L., Cadrobbi, P., Frauca, E., Camarena, C. et al. (1998). Outcome of chronic hepatitis B in caucasian children during a 20-year observation period. *J Hepatol,* (29), 184–190.

Boxall, E.H., Sira, J., Ballard, A.L., Davies, P., & Kelly, D.A. (2006). Long term follow up of hepatitis B carrier children treated with interferon and prednisolone. *J Med Virol,* (78), 888–895.

Boxall, E.H., Sira, J., El-Shukri, N., Kelly, D.A. (2004a). Long-term persistence of immunity to hepatitis B after vaccination during infancy in a country where endemicity is low. *J Infect Dis,* 190(7), 1264–1269.

Boxall, E.H., Sira, J., Standish, R.A., Davies, P., Sleight, E., Dhillon, A.P. et al. (2004b). Natural history of hepatitis B in perinatally infected carriers. *Arch Dis Child Fetal Neonatal Ed,* 89(5), 456–460.

Chu, C.M. (2000). Natural history of chronic hepatitis B virus infection in adults, emphasis on the occurrence of cirrhosis and hepatocellular carcinoma. *J Gastroenterol Hepatol,* 15(Suppl), E25–E30.

D'Antiga, L., Aw, M., Atkins, M. et al. (2006). Combined lamivudine/interferon alfa treatment in immune tolerant children perinatally infected with Hepatitis B; a pilot study. *J Pediatr,* (148), 228–233.

European Association for the Study. (2009). EASL clinical practice guidelines: management of hepatitis B. *J Hepatol,* 50(2), 227–242.

Ghany, M. & Liang, T.J. (2007). Drug targets and molecular mechanisms of drug resistance in chronic Hepatitis B. *Gastroenterology,* (132), 1574–1585.

Hadziyannis, S., Tassopoulos, N.C., Heathcote, E.J., Chang, T.T., Kitis, G., Rizzetto, M. et al. Adefovir Dipivoxil 438 Study Group. (2005). Long-term therapy with Adefovir Dipivoxil 438 Study Group. *N Engl J Med,* (352), 2673–2681.

Jara, P., & Bortolotti, F. (1999). Interferon-α treatment of chronic hepatitis B in childhood: a consensus advice based on experience in European children. *J Pediatr Gastroenterol Nutr,* (29), 163–170.

Jonas, M., Kelly, D., & Mizerski, J. (2002). Clinical trial of lamivudine in children with chronic hepatitis B. *N Engl J,* (346), 1706–1713.

Jonas, M.M., Kelly, D., Pollack, H., Mizerski, J., Sorbel, J., Frederick, D. et al. (2008). Safety, efficacy, and pharmacokinetics of adefovir dipivoxil in children and adolescents (age 2 to <18 years) with chronic hepatitis B. *Hepatology,* 47(6), 1863–1871.

Lai, C.L., Leung, N., Teo, E.K. et al. (2005). A 1-year trial of telbivudine, lamivudine and the combination in patients with hepatitis B e antigen-positive chronic hepatitis B. *Gastroenterology,* (129), 528–536.

Lau, G.K., Piratvisuth, T., Luo, K.X. et al. (2005). Peginterferon Alfa2a, lamivudine, and the combination for HBeAG-positive chronic hepatitis B. *N Engl J Med,* (352), 2682–2695.

Lavanchy, D. (2004). Hepatitis B virus epidemiology, disease burden, treatment and current and emerging prevention and control measures. *J Viral Hepat,* (11), 97–107.

Lee, W.S., McKiernan, P., & Kelly, D.A. (2005). Etiology, outcome and prognostic indicators of childhood fulminant hepatic failure in the United Kingdom. *JPGN,* (40), 575–581.

Lok, A.S. & McMahon, B.J. (2007). Chronic hepatitis B. *Hepatology,* (45), 507–539.

Moore, S.W., Millar, A.J., Hadley, G.P., Ionescu, G., Kruger, M., Poole, J. et al. (2004). Hepatocellular carcinoma and liver tumors in South African children: a case for increased prevalence. *Cancer,* 101(3), 642–649.

Perillo, R., Schiff, E., Magill, A., & Murray, A. (1999) In vivo demonstration of sensitivity of YMDD variants to adefovir. *Gastroenterol,* (116), A1261.

Petersen, K.M., Bulkow, L.R., McMahon, M.D., Zanis, C., Getty, M., Peters, H., & Parkinson, A.J. (2004). Duration of hepatitis B immunity in low risk children receiving hepatitis B vaccinations from birth. *Pediatr Infect Dis J,* (23), 650–655.

Rehermann, B. (2003). Immune responses in hepatitis B virus infection. *Semin Liver Dis,* (23), 21–38.

Sherman, M., Yurdaydin, C., Sollano, J., Liaw, Y.F. et al. (2006). Entecavir for treatment of lamivudine-refractory HbeAg-positive chronic hepatitis B. *Gastroenterology,* 130(7), 2039–2049.

Sokal, E.M., Kelly, D.A., Mizerski, J., Badia, I.B., Areias, J.A., Schwarz, K.B. et al. (2006). Long term lamivudine therapy for children with HBeAg-positive chronic hepatitis B. *Hepatology,* 43(2), 225–232.

Terrault, N.A. (2002). Treatment of recurrent hepatitis B infection in liver transplant recipients. *Liver Trans,* 8(Suppl 10), S74–S81.

van Bommel, F., Zollner, B., Sarrazin, C. et al. (2006). Tenofovir for patients with lamivudine-resistance hepatitis B virus (HBV) infection and high HBV DNA level during adefovir therapy. *Hepatology,* 44(2), 318–325.

Treatment of Neonatal Fungal Infections

Cassandra Moran and Danny Benjamin

1 Introduction

Invasive candidiasis is a leading cause of infection-related morbidity and mortality in premature infants (Stoll et al., 2002; Kremer et al., 1992; Lee et al., 1998; Mittal et al., 1998; Friedman et al., 2000; Benjamin et al., 2004c). Risk factors for candidiasis include the following: extreme prematurity, use of antibiotics, parenteral nutrition, and postnatal steroids (Kaufman, 2003), and center differences in the incidence of candidiasis. Several studies have reported a 20% mortality rate, despite antifungal treatment (Stoll et al., 1996, 2002; Benjamin et al., 2004a). Antifungal therapy failure is correlated with birthweight: for those with a birthweight 751–1000 g the failure rate is 14%, and for those with a birthweight of 400–750 g, it is 31% (Benjamin et al., 2000, 2003b, 2006). While much progress has been made in the management of neonatal candidiasis, definitive guidelines for prophylaxis and treatment are limited (Benjamin et al., 2003c). Routine prophylaxis is not recommended for neonates and use of either fluconazole or amphotericin B deoxycholate is considered 1st line therapy for treatment of invasive disease. However, new therapeutic options that may be safer and more efficacious are on the horizon.

2 Epidemiology

Candida is the third most common pathogen isolated in nosocomial blood stream infections in premature infants (Stoll et al., 1996; Kaufman, 2003; Benjamin et al., 2004a). Candidemia is diagnosed in approximately 3000 infants each year in the US and it mostly affects infants born <28 weeks gestational age. Mean age at onset ranges from 7 to 33 days, depending on weight and gestational age. Low

D. Benjamin (✉)
Division of Quantitative Sciences, Duke Clinical Research Institute, Duke University Pediatrics, Durham, NC 27715, USA
e-mail: danny.benjamin@duke.edu

A. Finn et al. (eds.), *Hot Topics in Infection and Immunity in Children VI*, Advances in Experimental Medicine and Biology 659, DOI 10.1007/978-1-4419-0981-7_11,
© Springer Science+Business Media, LLC 2010

birth weight is strongly associated with candidiasis. A prospective analysis, which included 6 Neonatal Intensive Care Units (NICUs) during a 3 year period, reported an incidence of 0.26% (3/1139) in infants weighing 2500 g or more (Saiman et al., 2000). However, in very low birth weight (VLBW, <1500 g birth weight) neonates, the incidence was 3.1%, and, for ELBW infants, up to 5.5%. Similarly, the Neonatal Research Network reported a cumulative incidence of 7% in ELBW infants. The Pediatrix group (over 100 nurseries reported similar rates) (Benjamin et al., 2003b, 2004c).

In addition to age and birth weight, the use of broad spectrum antibiotics, especially third-generation cephalosporins, is a well known risk factor. Such antibiotics promote fungal colonization by inhibition of competing bacterial growth (Kaufman and Fairchild, 2004). One multicenter cohort study reported that the use of a third-generation cephalosporin or carbapenem within 7 days prior to diagnosis was associated with neonatal candidiasis (Benjamin et al., 2003a). Other risk factors for invasive disease include histamine-2 blockers, steroids, abdominal surgery, mucoepithelial infection, central vascular catheters, hyperglycemia, total parental nutrition, and intralipids (Baley et al., 1986; Rowen et al., 1994; Botas et al., 1995; Huang et al., 1998; Benjamin et al., 2003a, 2005; Saiman et al., 2000; Makhoul et al., 2001; Kaufman, 2003).

Several studies have also indicated that colonization with *Candida* is associated with increased rates of invasive disease (Baley et al., 1986; Rowen et al., 1994; Huang et al., 1998). Colonization with *Candida* is more prevalent in early gestational age neonates and low birth weight infants, occurring in approximately 30% of infants <1500 g birth weight (Saiman et al., 2001). Although studies have identified colonization as a risk factor, a large prospective multi-center study demonstrated that colonization is not an independent risk factor for later development of disseminated candidiasis (Saiman et al., 2000).

2.1 Candida Species

Although over 200 known species of *Candida* exist (Dismukes et al., 2003), two account for 90% of neonatal candidiasis; they are the following: *C. albicans and C. parapsilosis* (Kossoff et al., 1998; Kim et al., 2003; Lopez Sastre et al., 2003; Roilides et al., 2004). *C. albicans* is more virulent than other non-*albicans Candida* with a higher morbidity and mortality compared to other species (Benjamin et al., 2003c, 2004c). *Candida parapsilosis* is now the most common *Candida* species isolated in some NICUs (Benjamin et al., 2000; Roilides et al., 2004). *C. parapsilosis* is usually associated with a lower mortality relative to other *Candida* species (Stoll et al., 2002; Benjamin et al., 2004c; Roilides et al., 2004). *Candida glabrata* is much less common as compared to adult patients (Fairchild et al., 2002) and is often resistant to fluconazole. *Candida krusei* and *C. lusitaniae* are also less common in neonates but are concerning because of their resistance to fluconazole and amphotericin B, respectively.

3 Invasive Candidiasis

Candidemia can lead to systemic disease with end-organ damage. As expected, prolonged candidemia is a risk factor for systemic disease (Chapman and Faix, 2000); however, multi-organ disease can occur with one positive blood culture. Fortunately, the incidence of end-organ involvement has decreased because of early diagnosis, improved awareness regarding the pathogenicity of *Candida,* prompt removal of central lines, and earlier antifungal therapy (Benjamin et al., 2003c). Signs of candidiasis in the neonate are often non-specific and include temperature instability, apnea, hypotension, respiratory failure, need for mechanical ventilation, abdominal distension, and poor feeding (Fanaroff et al., 1998; Benjamin et al., 2000). Clinical assessment includes urine and CSF cultures and imaging of the CNS, abdomen, and heart (Benjamin et al., 2003c).

Organs that may be affected in invasive candidiasis include the skin, brain, kidneys, and eyes. Congenital cutaneous candidiasis is an uncommon infection that typically presents in the first day of life as a vesicular or "burn-like" rash. It is often associated with positive blood and CSF cultures in VLBW neonates (Darmstadt et al., 2000). *Candida* in preterm infants may also cause invasive fungal dermatitis. These lesions develop after birth but within the first 2 weeks of life (Rowen, 2003). Similarly, these skin lesions are associated with positive blood cultures. *Candida* infections of the central nervous system (CNS) may present as granulomas, vasculitis, or abscesses (Faix and Chapman, 2003) and usually have normal cerebral spinal fluid (CSF) parameters. One study reported that only 25% of neonates with positive CSF cultures had abnormal CSF parameters (Lee et al., 1998). *Candida* involvement of the urinary tract may present as isolated candiduria "fungus balls" within the renal pelvis and other urinary structures or with invasion of the renal parenchyma. Systemic antifungal therapy is typically adequate for treatment provided there is no urinary tract obstruction (Benjamin et al., 1999; Bryant et al., 1999). Surgery is rarely needed. Endophthalmitis is a less common complication compared to kidney or CNS disease (Benjamin et al., 2003c). However, because candidemia without endophthalmitis has been associated with an increased risk of severe retinopathy of prematurity (Friedman et al., 2000), an ophthalmologic exam following candidemia is recommended.

4 End Organ Assessment

Morbidity from neonatal candidiasis is significant. Neurodevelopmental disability, severe retinopathy of prematurity, periventricular leukomalacia, and chronic lung disease are well described complications associated with disseminated candidiasis (Lee et al., 1998) (Kremer et al., 1992; Mittal et al., 1998; Friedman et al., 2000). Additionally, associations between candidiasis and intraventricular hemorrhage have been seen by some investigators (Lee et al., 1998) although this is not a consistent observation by other groups (Friedman et al., 2000).

Neurodevelopmental impairment after candidiasis can be assessed by a composite endpoint defined by one or more of the following: (a). Bayley Scales of Infant Development score <70 on the mental developmental index (MDI), (b). Bayley score <70 on the psychomotor developmental index (PDI), (c). Cerebral palsy (either moderate or severe), (d). Blindness, (e). Deafness.

Benjamin et al. evaluated neurodevelopmental assessments in 3049 infants with a history of candidiasis. The follow-up data performed at 18–22 months of age showed that median Bayley MDI and PDI scores were lower for those with a history of *Candida* (P <.001). In fact, infants were more likely to have Bayley MDI or PDI scores of <70, moderate or severe cerebral palsy as well as blindness and deafness when compared to neonates without a known *Candida* infection.

5 Treatment

Prompt removal or replacement of a central venous catheter is critical for prompt, effective therapy. Delayed catheter removal is associated with increased mortality and worse neurodevelopmental outcome (Eppes et al., 1989; Karlowicz et al., 2000; Benjamin et al., 2004c). Delayed removal/replacement of the central catheter following candidemia has been shown to be associated with worse short-term and long-term outcomes in a cohort of 320 premature infants with candidemia: in those with delayed removal/replacement, mortality rates were 37 vs. 21% in prompt removal/replacement; neurodevelopmental impairment (NDI) among survivors: 63% in delayed vs. 45% in prompt; time to clear Candida from the blood: 7.3 days in delayed vs. 5.1 in prompt removal. The association between death/NDI and delayed catheter removal remained strong in multivariable analysis: $OR = 2.7$ 95%CI (1.3, 5.8) (Benjamin et al., 2006).

Microbiologic clearance of CSF, urine, and blood should be documented; however, despite the presence of Candida, CSF WBC and culture can be normal. Culture results will guide therapy if persistently positive. Additionally, urine culture results may guide therapy with respect to lipid complex products. Two negative blood cultures separated by at least 24 h should be obtained. Because central nervous system disease is difficult to diagnose, is frequently observed in neonataol autopsy, and because candidiasis is so closely related to neurodevelopmental impairment, central nervous system disease should be assumed. An eye exam should be done, as the incidence of eye involvement is 1–3%, and it influences therapy because of the need for penetration into the vitreous. Echocardiogram should be completed because valvular disease will influence length of therapy. An abdominal ultrasound to evaluate renal abscesses (5% incidence) may influence the type of therapy.

Definitive antifungal therapy is less clear because of the lack of well-powered randomized trials. Many agents including new antifungal drugs with improved safety profiles are currently available; however, PK data in neonates are limited. First line agents include amphotericin B deoxycholate (1 mg/kg/day), fluconazole (12 mg/kg/day) and (if urine cultures are negative) lipid complex formulations

(5 mg/kg/day). Micafungin kinetics are known, and 10 mg/kg/day of this product may be used. The current formulation of anidulafungin contains alcohol and the kinetics are not known in children under 2 years of age. The dosing information of caspofungin in young infants is sparse, and the dosing of posaconazole and voriconazole are unknown in the young infant.

5.1 Amphotericin B deoxycholate

Amphotericin B deoxycholate binds to ergosterol, a component of the fungal cell membrane, leading to enhanced cell permeability and cell death. As most *Candida* species, except *C. lusitaniae* and occasionally *C. glabrata* and *C. krusei,* are susceptible to amphotericin B, it is the most commonly used antifungal for the treatment of neonatal candidiasis (Rowen and Tate, 1998), (Minari et al., 2001). However, despite such broad use, safety and efficacy data in neonates are limited. Furthermore, pharmacokinetic variability in neonates and a longer half life as well as improved CF penetration relative to adults have been reported (Baley et al., 1990). The recommended dose is 1.0 mg/kg intravenously given every 24 h (Baley et al., 1990; Serra et al., 1991). Test doses are not needed in neonates (Juster-Reicher et al., 2003). Nephrotoxicity is the main serious side effect (Baley et al., 1984). However, amphotericin B deoxycholate is thought to be well tolerated in premature infants (Kingo et al., 1997; Linder et al., 2003). Electrolyte abnormalities, however, (particulary hypokalemia) (Linder et al., 2003), are not uncommmon. Rigors, fever, and chills are not generally seen in neonates.

5.2 Lipid Preparations of Amphotericin B

The main advantage of lipid-based formulations of amphotericin B is the decreased renal toxicity with relatively higher doses of the parent drug (Juster-Reicher et al., 2003). The decrease in renal toxicity is likely due to less drug penetration of the kidneys (Bekersky et al., 2002). Lipid-based formulations of amphotericin B are used when patients have renal insufficiency or when candidemia persists despite standard doses of amphotericin B deoxycholate (Linder et al., 2003) and multiple preparations are available. One controlled trial evaluated and compared the effectiveness of the liposomal preparations on fifty-six neonates. Each neonate was given one of three amphotericin products: amphotericin B deoxycholate ($n = 34$) or if their serum creatinine was >1.2 mg/dL they were given either L-am B ($n = 6$) or ABCD ($n = 16$) (Linder et al., 2003). The overall mortality rate was 14.8% (8/56). Mortality was similar for the three groups [amphotericin B deoxycholate 14.7% (5/34), L-amB 16.7% (1/6), and ABCD 12.5% (2/16)]; however, due to inadequate numbers, the trial could not compare efficacy. No renal toxicity was seen during therapy for any group.

5.3 5-Fluorocytosine (Flucytosine, 5-FC).

Flucytosine is a fluorinated pyrimidine analogue that inhibits fungal DNA synthesis and interferes with RNA synthesis. It can be used in combination with amphotericin B to treat central nervous system *Candida* infections. However, because the time to clear CSF is longer in neonates with Candida meningitis who are treated with flucytosine plus amphotericin B when compared to amphotericin B alon, its benefit in this regard has been questioned (Benjamin et al., 2006). Due to the rapid drug resistance acquired when used alone, it should not be used as monotherapy (Bennett, 1996). The recommended dosage is 100–150 mg/kg/day divided every 6 h. The main side effect is bone marrow suppression, which can include agranulocytosis and aplastic anemia, as seen with drug levels >100 μg/mL. Therefore, levels should be followed in neonates, especially those with impairment in renal excretion (Frattarelli et al., 2004).

The benefits of flucytosine in premature infants must be carefully weighed with the risks of side effects that affect the bone marrow, hepatic, and renal systems. CNS infections can be cleared successfully without flucytosine. When the decision is made to treat, flucy to sine levels should be monitored closely.

5.4 The Azoles

Fluconazole, available in both oral and intravenous formulations, is the most commonly used azole antifungal in neonates to treat candidiasis (Rowen and Tate, 1998; Wenzl et al., 1998). Fluconazole should be given 12 mg/kg/day to young infants (Wade et al., 2008). Oral absorption is nearly 100%. Fluconazole is fungistatic; it penetrates the CSF, kidney, and liver. In neonates, the most common side effect is mild elevation of liver enzymes (Fasano et al., 1994).

In a randomized trial of 23 neonates, one study compared fluconazole (10 mg/kg initial dose then 5 mg/kg/day) to amphotericin B deoxycholate (1 mg/kg/day) (Driessen et al., 1996). The mortality rate was 45.5% (5/11) for those who received amphotericin B, and 33% (4/12) for those who received fluconazole groups. Another study of 40 neonates given parenteral fluconazole (6 mg/kg/day for 6–48 days) for candidiasis reported an attributable mortality rate of 10% (4/40) and an overall mortality of 20% (8/40). (Huttova et al., 1998). The majority of patients had *Candida albicans*. Although two patients experienced a 5-fold elevation in liver enzymes, and two with elevated serum creatinine levels, discontinuation of therapy was not necessary. Because clearance increases with post-natal age, to achieve systemic drug exposures recommended for older children and adults, dosing of at least 12 mg/kg Q24 is appropriate in young infants with candidiasis. If <30 weeks EGA *and* <2 weeks PNA *and* creatinine >1, then give first dose at 12 mg/kg (loading), and then follow creatinine; if creatinine stays >1, then consider 6 mg/kg/day, if creatinine drops <1, then continue 12 mg/kg/day.

C. albicans (MIC_{90}–1 μg/mL) and *C. parapsilosis* (MIC_{90}–2 μg/mL) the two most commonly identified *Candida* species in neonates, are sensitive to fluconazole (Pfaller et al., 1999). However, two non-*albicans* species, *C. glabrata* and *C. krusei,* are usually resistant (MIC_{90}–64 μg/mL) (Rowen et al., 1999), due to alterations in the target enzyme, sterol 14-α-demethylase or by enhanced efflux out of the fungal cell (Marichal et al., 1999; Sanglard, 2002).

Voriconazole, a second-generation triazole, is active in vitro against most *Candida* species including *C. glabrata* and *C. krusei* (Muller et al., 2000). Side effects include torsades de pointes, allergic reactions, elevated transaminases, and visual disturbances (Walsh et al., 2001). A study of 58 pediatric patients receiving voriconazole reported elevated transaminases in 13.8% (8/58), abnormal vision in 5.2% (3/58), and photosensitivity in 5.2% (3/58), (Walsh et al., 2002). Given the visual side effects and its unknown impact on the developing retina, neonates who are predisposed for retinopathy of prematurity may not be good candidates for this agent. Pharmacokinetic data in neonates is unknown but for aspergillosis it is the treatment of choice.

5.5 *The Echinocandins*

The echinocandins include caspofungin, micafungin, and anidulafungin. The echinocandins are active against all *Candida* species (Bartizal et al., 1997) and are available for intravenous administration (Frattarelli et al., 2004). The mechanism of action is inhibition of 1,3 β-D-glucan synthase (Walsh et al., 2000). Little is known about the appropriate dosing of caspofungin in young infants, although it has been used (Odio et al., 2004). Side effects include hypokalemia, elevated transaminases, and anemia. Anidulafungin exhibits linear kinetics in children 2–17 years of age, (Benjamin et al., 2004b), and studies are in progress to define the PK in toddlers and infants using a new formulation without alcohol as a vehicle.

Micafungin has been evaluated in the neonates in three NICHD-sponsored Pediatric Pharmacology Research Network studies. Dosages of 1.5 mg/kg, 3 mg/kg (Heresi et al., 2003), 7 mg/kg, 10 mg/kg (Benjamin), and 15 mg/kg (Smith) have been studied. Premature infants have a faster clearance and a need for central nervous system penetration. Echinocandins have a dose-response relationship; therefore, the one echinocandin for which firm dosing recommendations can be made is micafungin. This product should be given at 10–12 mg/kg/day.

References

Baley, J.E., Kliegman, R.M. et al. (1984). Disseminated fungal infections in very low-birth-weight infants: therapeutic toxicity. *Pediatrics,* 73(2), 153–157.

Baley, J.E., Kliegman, R.M. et al. (1986). Fungal colonization in the very low birth weight infant. *Pediatrics,* 78(2), 225–232.

Baley, J.E., Meyers, C. et al. (1990). Pharmacokinetics, outcome of treatment, and toxic effects of amphotericin B and 5-fluorocytosine in neonates. *J Pediatr,* 116(5), 791–797.

Bartizal, K., Gill, C.J. et al. (1997). In vitro preclinical evaluation studies with the echinocandin antifungal MK-0991 (L-743,872). *Antimicrob Agents Chemother,* 41(11), 2326–2332.

Bekersky, I., Fielding, R.M. et al. (2002). Pharmacokinetics, excretion, and mass balance of liposomal amphotericin B (AmBisome). and amphotericin B deoxycholate in humans. *Antimicrob Agents Chemother,* 46(3), 828–833.

Benjamin, D.K., Jr., DeLong, E.R. et al. (2003a). Empirical therapy for neonatal candidemia in very low birth weight infants. *Pediatrics,* 112(3 Pt 1), 543–547.

Benjamin, D.K., DeLong, E. et al. (2004a). Mortality following blood culture in premature infants: increased with gram-negative bacteremia and candidemia, but not gram-positive bacteremia. *J Perinatol,* 24(3), 175–180.

Benjamin, D.K., Driscoll, T. et al. (2004b). *Safety and Pharmacokinetics of Anidulafungin in Pediatric Patients with Neutropenia.* Interscience Conference on Antimicrobial Agents and Chemotherapy, Washington, D.C.

Benjamin, D.K., Jr., Fisher, R.G. et al. (1999). Candidal mycetoma in the neonatal kidney. *Pediatrics,* 104(5 Pt 1), 1126–1129.

Benjamin, D.K., Jr., Garges, H. et al. (2003b). Candida bloodstream infection in neonates. *Semin Perinatol,* 27(5), 375–383.

Benjamin, D.K., Jr., Poole, C. et al. (2003c). Neonatal candidemia and end-organ damage: a critical appraisal of the literature using meta-analytic techniques. *Pediatrics,* 112(3 Pt 1), 634–640.

Benjamin, D.K., Jr., Ross, K. et al. (2000). When to suspect fungal infection in neonates: a clinical comparison of Candida albicans and Candida parapsilosis fungemia with coagulase-negative staphylococcal bacteremia. *Pediatrics,* 106(4), 712–718.

Benjamin, D.K., Jr., Stoll, B.J. et al. (2004c). *Neonatal Candidiasis Among Infants <1000 g Birthweight: Risk Factors, Mortality, and Neuro-Developmental Outcomes at 18–22 months.* Interscience Conference on Antimicrobial Agents and Chemotherapy.

Benjamin, D.K., Jr., Stoll, B.J. et al. (2005). Neonatal candidiasis among extremely low birth weight infants: risk factors, mortality, and neuro-developmental outcomes at 18–22 months. *Pediatrics,* 116(2), e241–e246.

Benjamin, D.K., Jr., Stoll, B.J. et al. (2006). Neonatal candidiasis among extremely low birth weight infants: risk factors, mortality rates, and neurodevelopmental outcomes at 18 to 22 months. *Pediatrics,* 117(1), 84–92.

Bennett, J. (1996). *The Pharmacological Basis of Therapeutics.* New York: McGraw-Hill.

Botas, C.M., Kurlat, I. et al. (1995). Disseminated candidal infections and intravenous hydrocortisone in preterm infants. *Pediatrics* 95(6), 883–887.

Bryant, K., Maxfield, C. et al. (1999). Renal candidiasis in neonates with candiduria. *Pediatr Infect Dis J,* 18(11), 959–963.

Chapman, R.L. & Faix, R.G. (2000). Persistently positive cultures and outcome in invasive neonatal candidiasis. *Pediatr Infect Dis J,* 19(9), 822–827.

Darmstadt, G.L., Dinulos, J.G. et al. (2000). Congenital cutaneous candidiasis: clinical presentation, pathogenesis, and management guidelines. *Pediatrics,* 105(2), 438–444.

Dismukes, W., Pappas, P. et al. (2003). *Clinical Mycology.* New York: Oxford University Press.

Driessen, M., Ellis, J.B. et al. (1996). Fluconazole vs. amphotericin B for the treatment of neonatal fungal septicemia: a prospective randomized trial. *Pediatr Infect Dis J,* 15(12), 1107–1112.

Eppes, S.C., Troutman, J.L. et al. (1989). Outcome of treatment of candidemia in children whose central catheters were removed or retained. *Pediatr Infect Dis J,* 8(2), 99–104.

Fairchild, K.D., Tomkoria, S. et al. (2002). Neonatal Candida glabrata sepsis: clinical and laboratory features compared with other Candida species. *Pediatr Infect Dis J,* 21(1), 39–43.

Faix, R.G. & Chapman, R.L. (2003). Central nervous system candidiasis in the high-risk neonate. *Semin Perinatol,* 27(5), 384–392.

Fanaroff, A.A., Korones, S.B. et al. (1998). Incidence, presenting features, risk factors and significance of late onset septicemia in very low birth weight infants. The National Institute of Child Health and Human Development Neonatal Research Network. *Pediatr Infect Dis J,* 17(7), 593–598.

Fasano, C., O'Keeffe, J. et al. (1994). Fluconazole treatment of neonates and infants with severe fungal infections not treatable with conventional agents. *Eur J Clin Microbiol Infect Dis,* 13(4), 351–354.

Frattarelli, D.A., Reed, M.D. et al. (2004). Antifungals in systemic neonatal candidiasis. *Drugs,* 64(9), 949–968.

Friedman, S., Richardson, S.E. et al. (2000). Systemic Candida infection in extremely low birth weight infants: short term morbidity and long term neurodevelopmental outcome. *Pediatr Infect Dis J,* 19(6), 499–504.

Heresi, G., Gerstmann, D. et al. (2003). *A Pharmacokinetic Study of Micafungin (FK463) in Premature Infants (abstract).* Seattle, Washington: Pediatric Academic Society.

Huang, Y.C., Li, C.C. et al. (1998). Association of fungal colonization and invasive disease in very low birth weight infants. *Pediatr Infect Dis J,* 17(9), 819–822.

Huttova, M., Hartmanova, I. et al. (1998). Candida fungemia in neonates treated with fluconazole: report of forty cases, including eight with meningitis. *Pediatr Infect Dis J,* 17(11), 1012–1015.

Juster-Reicher, A., Flidel-Rimon, O. et al. (2003). High-dose liposomal amphotericin B in the therapy of systemic candidiasis in neonates. *Eur J Clin Microbiol Infect Dis,* 22(10), 603–607.

Karlowicz, M.G. (2003). Candidal renal and urinary tract infection in neonates. *Semin Perinatol,* 27(5), 393–400.

Karlowicz, M.G., Hashimoto, L.N. et al. (2000). Should central venous catheters be removed as soon as candidemia is detected in neonates? *Pediatrics,* 106(5), E63.

Kaufman, D. (2003). Strategies for prevention of neonatal invasive candidiasis. *Semin Perinatol,* 27(5), 414–424.

Kaufman, D. & Fairchild, K.D. (2004). Clinical microbiology of bacterial and fungal sepsis in very-low-birth-weight infants. *Clin Microbiol Rev,* 17(3), 638–680.

Kim, J.O., Garofalo, L. et al. (2003). Candida dubliniensis infections in a pediatric population: retrospective identification from clinical laboratory isolates of Candida albicans. *J Clin Microbiol,* 41(7), 3354–3357.

Kingo, A.R., Smyth, J.A. et al. (1997). Lack of evidence of amphotericin B toxicity in very low birth weight infants treated for systemic candidiasis. *Pediatr Infect Dis J,* 16(10), 1002–1003.

Kossoff, E.H., Buescher, E.S. et al. (1998). Candidemia in a neonatal intensive care unit: trends during fifteen years and clinical features of 111 cases. *Pediatr Infect Dis J,* 17(6), 504–508.

Kremer, I., Naor, N. et al. (1992). Systemic candidiasis in babies with retinopathy of prematurity. *Graefes Arch Clin Exp Ophthalmol,* 230(6), 592–594.

Lee, B.E., Cheung, P.Y. et al. (1998). Comparative study of mortality and morbidity in premature infants (birth weight, < 1,250 g) with candidemia or candidal meningitis. *Clin Infect Dis,* 27(3), 559–565.

Linder, N., Klinger, G. et al. (2003). Treatment of candidaemia in premature infants: comparison of three amphotericin B preparations. *J Antimicrob Chemother,* 52(4), 663–667.

Lopez Sastre, J.B., Coto Cotallo, G.D. et al. (2003). Neonatal invasive candidiasis: a prospective multicenter study of 118 cases. *Am J Perinatol,* 20(3), 153–163.

Makhoul, I.R., Kassis, I. et al. (2001). Review of 49 neonates with acquired fungal sepsis: further characterization. *Pediatrics,* 107(1), 61–66.

Marichal, P., Koymans, L. et al. (1999). Contribution of mutations in the cytochrome P450 14alpha-demethylase (Erg11p, Cyp51p) to azole resistance in Candida albicans. *Microbiology,* 145(Pt 10), 2701–2713.

Minari, A., Hachem, R. et al. (2001). Candida lusitaniae: a cause of breakthrough fungemia in cancer patients. *Clin Infect Dis,* 32(2), 186–190.

Mittal, M., Dhanireddy, R. et al. (1998). Candida sepsis and association with retinopathy of prematurity. *Pediatrics,* 101(4 Pt 1), 654–657.

Muller, F.M., Weig, M. et al. (2000). Azole cross-resistance to ketoconazole, fluconazole, itraconazole and voriconazole in clinical Candida albicans isolates from HIV-infected children with oropharyngeal candidosis. *J Antimicrob Chemother,* 46(2), 338–340.

Odio, C.M., Araya, R. et al. (2004). Caspofungin therapy of neonates with invasive candidiasis. *Pediatr Infect Dis J*, 23(12), 1093–1097.

Pfaller, M.A., Messer, S.A. et al. (1999). Trends in species distribution and susceptibility to fluconazole among blood stream isolates of Candida species in the United States. *Diagn Microbiol Infect Dis*, 33(4), 217–222.

Roilides, E., Farmaki, E. et al. (2004). Neonatal candidiasis: analysis of epidemiology, drug susceptibility, and molecular typing of causative isolates. *Eur J Clin Microbiol Infect Dis*, 23(10), 745–750.

Rowen, J.L. (2003). Mucocutaneous candidiasis. *Semin Perinatol*, 27(5), 406–413.

Rowen, J.L., Rench, M.A. et al. (1994). Endotracheal colonization with Candida enhances risk of systemic candidiasis in very low birth weight neonates. *J Pediatr*, 124(5 Pt 1), 789–794.

Rowen, J.L. & Tate, J.M. (1998). Management of neonatal candidiasis. Neonatal Candidiasis Study Group. *Pediatr Infect Dis J*, 17(11), 1007–1011.

Rowen, J.L., Tate, J.M. et al. (1999). Candida isolates from neonates: frequency of misidentification and reduced fluconazole susceptibility. *J Clin Microbiol*, 37(11), 3735–3737.

Saiman, L., Ludington, E. et al. (2000). Risk factors for candidemia in neonatal intensive care unit patients. The National Epidemiology of Mycosis Survey Study Group. *Pediatr Infect Dis J*, 19(4), 319–324.

Saiman, L., Ludington, E. et al. (2001). Risk factors for Candida species colonization of neonatal intensive care unit patients. *Pediatr Infect Dis J*, 20(12), 1119–1124.

Sanglard, D. (2002).Clinical relevance of mechanisms of antifungal drug resistance in yeasts. *Enferm Infecc Microbiol Clin*, 20(9); 462–469, quiz 470, 479.

Saxen, H., Hoppu, K. et al. (1993). Pharmacokinetics of fluconazole in very low birth weight infants during the first two weeks of life. *Clin Pharmacol Ther*, 54(3), 269–277.

Serra, G., Mezzano, P. et al. (1991). Therapeutic treatment of systemic candidiasis in newborns. *J Chemother*, 3(Suppl 1), 240–244.

Smego, R.A., Jr., Perfect, J.R. et al. (1984). Combined therapy with amphotericin B and 5-fluorocytosine for Candida meningitis. *Rev Infect Dis*, 6(6), 791–801.

Stoll, B.J., Gordon, T. et al. (1996). Late-onset sepsis in very low birth weight neonates: a report from the National Institute of Child Health and Human Development Neonatal Research Network. *J Pediatr*, 129(1), 63–71.

Stoll, B.J., Hansen, N. et al. (2002). Late-onset sepsis in very low birth weight neonates: the experience of the NICHD Neonatal Research Network. *Pediatrics*, 110(2 Pt 1), 285–291.

Wade, K.C., Wu, D., Kaufman, D.A., Ward, R.M., Benjamin, D.K., Jr., Sullivan, J.E. et al. On behalf of the NICHD Pediatric Pharmacology Research Unit. (2008). Population Pharmacokinetics of Fluconazole in Young Infants. *Antimicrob Agents Chemother*, (52), 4043–4049.

Walsh, T.J., Arguedas, A. et al. (2001). *Pharmacokinetics of Intravenous Voriconazole in Children after Single and Multiple Dose Administration*. 41st Interscience Conference on Antimicrobial Agents and Chemotherapy (abstract).

Walsh, T.J., Lutsar, I. et al. (2002). Voriconazole in the treatment of aspergillosis, scedosporiosis and other invasive fungal infections in children. *Pediatr Infect Dis J*, 21(3), 240–248.

Walsh, T.J., Viviani, M.A. et al. (2000). New targets and delivery systems for antifungal therapy. *Med Mycol*, 38(Suppl 1), 335–347.

Wenzl, T.G., Schefels, J. et al. (1998). Pharmacokinetics of oral fluconazole in premature infants. *Eur J Pediatr*, 157(8), 661–662.

Malaria in Pregnancy and the Newborn

Stephen J. Rogerson

1 The Global Burden of Malaria

Each year, 40% of the world's population is exposed to the risk of malaria infection. Approximately 500 million people suffer clinical disease episodes of malaria, and around one million die from it. The greater part of the world's malaria burden falls on Africa, but recent analyses suggest the amount of malaria in Asia has been underestimated (Snow et al., 2005). Five *Plasmodium* species infect humans: *P. falciparum*, *P. vivax*, *P. malariae*, *P. ovale*, and *P. knowlesi*. The last, which is a common parasite of monkeys, has only recently been described as a human pathogen (Singh et al., 2004), but appears to be quite widespread in South East Asia. The great majority of severe disease episodes and deaths are due to *P. falciparum*, but it is becoming increasingly clear that *P. vivax* can also cause severe disease episodes and deaths (Genton et al., 2008; Tjitra et al., 2008). The main presentations of severe and life-threatening malaria are severe anaemia, cerebral malaria (unrouseable coma associated with malaria infection) and respiratory distress. Most deaths occur in young children, but pregnant women are also at particularly high risk, especially when they have lower malaria immunity.

2 The Burden of Malaria in Pregnancy

Each year, over 50 million pregnancies occur in malaria-endemic areas, and many pregnant women suffer effects of malaria in pregnancy. In Africa, around one in four women have placental malaria infection at delivery (Desai et al., 2007). Malaria contributes to maternal deaths, most often due to maternal anaemia. Estimates suggest that 10,000 women die, each year, from severe anaemia due to malaria (Guyatt and Snow, 2001). Low birth weight (LBW, < 2500 g) due to malaria is even more common, with an estimated 600,000 babies born with LBW each year, and

S.J. Rogerson (✉)
Department of Medicine (RMH/WH), Post Office Royal Melbourne Hospital, Parkville, VIC 3050, Australia
e-mail: sroger@unimelb.edu.au

A. Finn et al. (eds.), *Hot Topics in Infection and Immunity in Children VI*, Advances in Experimental Medicine and Biology 659, DOI 10.1007/978-1-4419-0981-7_12, © Springer Science+Business Media, LLC 2010

75,000–200,000 infant deaths ascribed to LBW consequent upon maternal malaria (Steketee et al., 2001). Moreover, the long term effects of these in utero insults due to malaria are unknown, and maternal malaria is likely to have effects on growth, development and even risk of adult-onset diseases (Barker, 2006).

3 Malaria Species in Pregnancy

Of the five species, *P. falciparum* is the most widely studied, is more common in pregnant than non-pregnant women, and is associated with adverse birth outcomes. *P. vivax*, the second most prevalent infection, has been associated with LBW and maternal anaemia in studies from India, Thailand and Indonesia (Nosten et al., 1999; Poespoprodjo et al., 2008; Singh et al., 1999), but it is not clear whether it becomes more common or severe in pregnancy. Of the different species, only *P. falciparum* is known to sequester in the placenta, and this placental sequestration is believed to be central to many of the manifestations of falciparum malaria in pregnancy. For *P. ovale, P. malariae and P. knowlesi*, the risks and consequences of infection during pregnancy are unknown.

4 Clinical Malaria in Pregnancy

In semi-immune adults, clinical disease from malaria is rare, and it has been thought that pregnant women are similarly unlikely to be symptomatic from the infections they carry. Two recent studies and some indirect data suggest this may not be the case. Women in Mozambique and Ghana who were parasitaemic more frequently had fever and other malaria symptoms, such as headache, dizziness and fatigue than matched, aparasitemic women, but these symptoms had a poor predictive value for malaria (Bardaji et al., 2008; Tagbor et al., 2008). In our Malawi studies, a history of febrile symptoms in the preceding week was strongly associated with placental malaria at delivery (odds ratio (OR) 5.8, 95% confidence interval (CI) 3.4–9.7, $p < 0.001$). Whilst it has been recognized that non-immune women frequently have symptoms (and sometimes severe disease) associated with malaria in pregnancy, these recent studies suggest such symptoms are not uncommon in high transmission areas, but warn that confirmation of infection is important to avoid inappropriate treatment with antimalarials.

5 Timing of Infection

Most cohort and cross-sectional studies of parasite prevalence have tested for parasitaemia in mid to late pregnancy, or at delivery (reviewed in Desai et al., 2007), in part because few women present in Africa to antenatal care before the second trimester of pregnancy. However, a few studies have managed to examine women in first or early second trimester for parasitaemia, and compare parasite prevalence across gestation. Interestingly, all studies show that the prevalence of infection is highest in the late first or early second trimester, and falls over gestation, and this seems to be the case regardless of transmission intensity and gravidity (Brabin, 1983; Brabin and Rogerson, 2001; Coulibaly et al., 2007). One possible explanation is that immunity to malaria develops over the course of the pregnancy, helping to suppress parasitaemia.

6 Susceptibility to Malaria in Pregnancy

There are a number of reasons why pregnant women may be at particular risk for malaria (Table 1). They are more attractive to mosquitoes than non-pregnant adults, probably because they exhale more carbon dioxide (Lindsay et al., 2000). The altered immunological and hormonal environment in pregnancy predisposes to a number of infectious diseases, such as listeriosis, CMV and hepatitis E (Hart, 1988). Malaria has been associated with elevated levels of corticosteroids in pregnancy (Vleugels et al., 1989), and other hormones have not been systematically studied. It has been postulated that changes in the Th1/Th2 cytokine balance in pregnancy predispose to malaria, while on the other hand malaria infection itself induces active Th1 and inflammatory cytokine responses (reviewed in Rogerson et al., 2007). One particularly important factor in the predisposition to malaria is the ability of *P. falciparum*-infected erythrocytes to sequester in the placenta. Such infected erythrocytes express a unique subset of variant surface antigens, or VSAs on their surface (Beeson et al., 1999; Maubert et al., 1999). As expression of these VSAs is restricted to pregnancy, a woman who is pregnant for her first time has a well-developed immunity to parasites expressing other VSAs, but lacks immunity to pregnancy-specific VSAs expressed by placental parasites, which exploit this "hole" in her existing immune response (Hviid, 2004).

Table 1 Reasons why pregnant women are at special risk of malaria

Increased susceptibility to mosquito bites
Altered cell mediated immunity in pregnancy
Hormonal changes associated with pregnancy
Placental sequestration of parasitized cells
Immune evasion by placental parasites

7 Gravidity and Age Influence Parasite Prevalence in Pregnancy

Women in their first pregnancy are at increased risk of malaria infection, and with subsequent pregnancies, their predisposition decreases (Desai et al., 2007). The intensity of malaria transmission may influence the rate of this decline (Fig. 1). In a high-transmission area of Malawi, pregnant women were highly likely to be parasitaemic at first antenatal clinic visit, and the rate fell quite steeply with

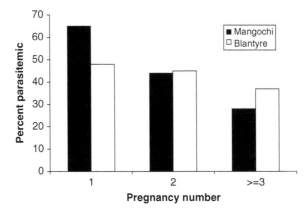

Fig. 1 Rates of malaria parasitaemia among pregnant women at first antenatal visit in Mangochi (*black*; high transmission) and Blantyre (*white*; moderate transmission), Malawi. Parasite rates decline with gravidity, and do so from a higher base and with greater rapidity in the higher transmission area. Adapted from Rogerson and Menendez, 2006 with permission of Expert Reviews Ltd

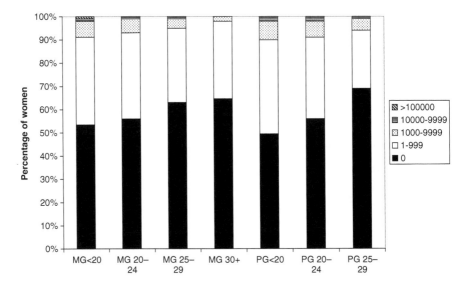

Fig. 2 Age is a co-determinant of susceptibility to malaria in pregnancy. At first antenatal visit, the proportion of multigravidae (MG) or primigravidae (PG) who were parasitaemic declined as maternal age increased. A parasitaemic women are shown in (*black*), women with low grade parasitaemia in (*white*), moderate in (*stippled*), high in (*grey*) and very high in (*hatched lines*). Moderate and high density parasitaemia (*stippled,grey* or *hatched* sections) was uncommon among older women

subsequent pregnancies. On the other hand, in Blantyre, lower malaria transmission was associated with lower parasite prevalence in first pregnancy, and relatively little decrease in prevalence with subsequent pregnancies, suggesting a slower acquisition of pregnancy-specific immunity.

Gravidity undoubtedly influences protection from malaria, and development of immunity to pregnancy-specific VSAs forms a key component of such protection. Some epidemiological evidence suggests that non-pregnancy-specific immunity may also be important. In Malawi, for example, we found that age was a more important predictor of parasitaemia at antenatal booking than gravidity (Fig. 2), and similar findings have been reported from Mozambique (Saute et al., 2002). While immunity that controls malaria parasitaemia generally develops over the course of childhood (Marsh and Kinyanjui, 2006), where malaria transmission is lower, women may still be developing immunity against blood-stage infection in early reproductive life. One important consequence of this is that adolescents in developing countries, who are at high risk of poor reproductive outcomes (Brabin, 2004), may also be particularly likely to suffer from malaria in pregnancy.

8 Modelling Placental Malaria In Vitro

In vitro models of sequestration of infected erythrocytes have been in use for some time. Chinese hamster ovary (CHO) cells express chondroitin sulphate A, and parasites "panned" on these cells adhered to CSA (Rogerson et al., 1995, Fig. 3). When chondroitin sulphate A from animal sources, or extracted from placenta, is spotted onto Petri dishes, parasitized cells from placenta adhere to the CSA (Fried et al., 2006; Fried and Duffy, 1996). Frozen sections of placenta, the BeWO cell line and in vitro-derived syncytiotrophoblast are also useful (Haase et al., 2006; Lucchi et al., 2006). By using these tools, it has been demonstrated that parasitized cells panned

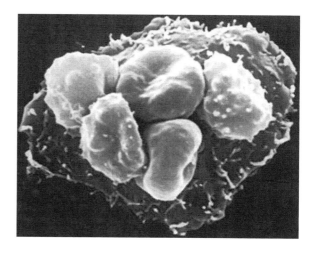

Fig. 3 Scanning electron micrograph of parasitised erythrocytes bound to Chinese hamster ovary cells

on CSA transcribe a *var* gene called *var2sca*, which encodes a protein on the surface of the infected red cell that mediates adhesion to CSA. This protein is the main target of the protective antibody response directed against the infected red cell surface (Duffy et al., 2005; Salanti et al., 2003). The parasite genome contains 60 *var* genes, each encoding a different red cell surface protein, but when *var2sca* is knocked out, parasites lose the ability to bind to CSA or to placenta (Duffy et al., 2006b; Viebig et al., 2005; Yosaatmadja et al., 2008). Parasites extracted from the placenta generally transcribe *var2sca* at high levels (Duffy et al., 2006a; Tuikue Ndam et al., 2005).

While expression of some other genes and proteins has been associated with placental malaria (Francis et al., 2007; Tuikue Ndam et al., 2008), the importance of these in sequestration or immunity is presently unclear.

Antibody immunity to parasite lines expressing the *var2csa* protein develops in a gender- and parity-dependent manner, and in some studies, levels of antibodies have been correlated with protection against adverse pregnancy outcomes (Duffy and Fried, 2003; Staalsoe et al., 2004). Because the *var2sca* gene is the most highly conserved of the *var* genes, and because it is an important target of immunity (Salanti et al., 2004), there is significant interest in developing a vaccine based on the *var2csa* protein. Identifying epitopes within the protein that are targets of protective immunity, and determining the degree of polymorphism within *var2csa* in isolates from different geographic regions are critical areas for future studies.

9 Placental Inflammation, Monocyte Infiltrates, and Pregnancy Outcomes

Clinically, LBW due to malaria results from a combination of fetal growth restriction and preterm delivery. In malaria-endemic Africa, the risk of LBW is approximately doubled by the presence of placental malaria, and this risk is highest in first pregnancy (Brabin et al., 1999; Guyatt and Snow, 2004). Here, the burden is principally that of fetal growth restriction, and malaria may be responsible for up to 70% of fetal growth restriction (Desai et al., 2007). Preterm delivery is more characteristically a feature of malaria in low-transmission settings, but even in Africa it may be responsible for over a third of premature deliveries (Desai et al., 2007). Given this substantial contribution to poor pregnancy outcomes, what have we learnt about the mechanisms underlying the development of LBW associated with malaria?

Placental histology studies have been very informative. Together with sequestration of infected erythrocytes in the placenta, there are often infiltrates of host leukocytes in the intervillous space (the maternal circulation of the placenta – Fig. 4). These cells are primarily monocytes and macrophages, with lymphocytes and neutrophils to a lesser extent, and many of these macrophages contain hemozoin, the by-product of the parasite's digestion of hemoglobin (Walter et al., 1982). Presence of hemozoin-laden macrophages correlates with low birth weight due to fetal growth restriction, while the density of infected red cells sequestered in the placenta correlates with pre-term delivery (Menendez et al., 2000; Rogerson et al.,

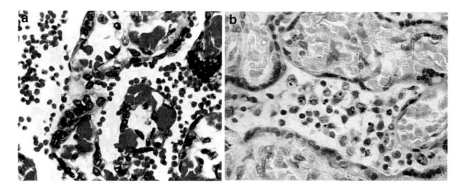

Fig. 4 Placental pathology. (**a**) dense parasite sequestration (**b**) dense monocyte infiltrates (**a**) haematoxylin and eosin (**b**) Giemsa; × 1000

2003b). This suggests two distinct pathophysiological mechanisms: acute events leading to premature parturition associated with dense placental parasitization, and more sub-acute or chronic events resulting in growth restriction.

Placental malaria is associated with release of inflammatory cytokines in the placenta. Increased levels of mRNA or protein of the cytokines tumour necrosis factor alpha, interferon gamma and interleukin 1 have been associated with malaria and low birth weight in different studies (Fried et al., 1998; Moormann et al., 1999; Rogerson et al., 2003a; Suguitan et al., 2003), and presumably, these cytokines affect the normal growth and functioning of the placenta. These studies do not yet reveal the events that interfere with fetal growth at the cellular level, or the exact mechanisms by which malaria can lead to premature parturition, but overall they suggest that placental accumulation of infected erythrocytes and inflammatory cells, together with products released by those inflammatory cells, probably have major effects on placental function, potentially interfering with successful pregnancy progression and fetal growth.

10 Treating Malaria in Pregnancy

The systematic exclusion of pregnant women from drug trials means that there is a profound lack of information regarding the safety, efficacy and dosing of antimalarials in pregnancy. The choice of drugs to treat malaria in pregnancy depends on the stage of the pregnancy, and the clinical state of the mother. When a woman is severely ill, our priority is to save her life, and the best available treatment is required. This means intravenous therapy with either quinine or artesunate (Rogerson and Menendez, 2006). Present data do not show either to have a clear advantage in pregnancy (Dondorp et al., 2005), and larger comparative studies are needed. Quinine is classed by the FDA as a Class C drug, but hard evidence that it is abortifacient is lacking (Phillips-Howard and Wood, 1996). Artesunate is associated with fetal resorption and congenital defects in rodents when used in early pregnancy

(Clark et al., 2004), but similar defects are not reported in humans (Ward, 2007). The pregnancy should be followed to term, and the outcome carefully documented.

For uncomplicated malarial illness, the choice of treatment depends on the trimester. During the first trimester, quinine, mefloquine and (for vivax malaria) chloroquine are all reportedly safe, whereas in the second and third trimesters sulphadoxine-pyrimethamine (SP) has been widely used. Artemisinins and artemisinin combination treatments, such as Coartem and dihydroartemisinin-piperaquine (Artekin) also can be used, with similar caveats about documenting pregnancy outcome to intravenous artesunate. For almost of these drugs there is a major lack of pharmacokinetic data; however, available data suggest dose modification may be required to achieve adequate therapeutic levels in pregnancy (Green et al., 2007; Ward et al., 2007). Such dose-modification studies have recently begun under the auspices of the Malaria in Pregnancy Consortium (www.mip-consortium.org).

11 Intermittent Preventive Treatment in Pregnancy

Intermittent preventive treatment in pregnancy, or IPTp, is the delivery of treatment doses of an antimalarial, at specified times, regardless of the presence of symptoms or infection. Based on studies from the 1980s and 1990s, the World Health Organization recommends at least two doses of an appropriate drug, at least a month apart, starting after quickening (when the mother detects fetal movements). A Cochrane review showed an increase in birth weight of 126 g among women receiving IPTp, compared to controls (Garner and Gulmezoglu, 2006). Presently, SP is the only proven safe effective drug in Africa, but neither the optimal number of doses, especially in women with HIV co-infection, who may require more than two doses (Parise et al., 1998), nor the appropriate dose for pregnant women (Green et al., 2007) are fully resolved. It is important that an IPTp policy using SP or other agents is properly implemented. When we compared women who went to ANC and did, or did not, receive IPTp doses, women who received the recommended two doses had a reduction of >50% in their prevalence of LBW babies, as well as reductions in anaemia and parasitaemia at delivery (Rogerson et al., 2000) (Fig. 5).

SP resistance is now common in much of Africa, and rates of parasitological treatment failure in children vary widely across the continent. Because pregnant women are semi-immune to malaria, it is less clear that SP resistance translates into loss of protection against pregnancy malaria, and a recent meta-analysis suggests that SP maintains its utility in pregnant women, up to moderately high levels of resistance (ter Kuile et al., 2007). If high-level resistance to SP emerges in Africa, as it did in Asia, it is probable that SP will have little role in the prevention of malaria in pregnancy, so a number of new drugs, including Artekin, mefloquine and artesunate, and SP plus azithromycin, are entering IPTp studies as part of the Malaria in Pregnancy Consortium's activities.

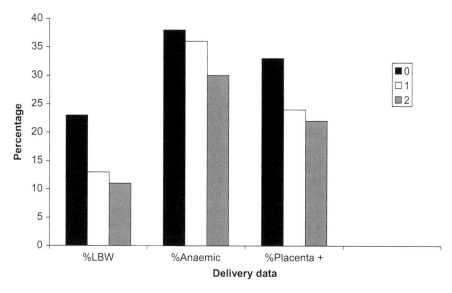

Fig. 5 Effectiveness of SP in Malawi, 1997–1999, by SP doses: 0, *black*; 1, *white*; 2 or more, *grey*. Among women who had attended antenatal clinic, and had opportunities to receive SP, those who were given 2 doses of SP had lower prevalences of low birth weight (%LBW), Hb < 11.0 g/dl (%anaemia) and placental parasitaemia (%Placenta +) than those who received 1 or 0 doses of SP

12 Malaria in the Newborn: Congenital Malaria

Congenital malaria is usually defined as peripheral blood parasites in the newborn within the first seven days of life – or longer, if no exposure to infected mosquitoes is possible (Menendez and Mayor, 2007). (The minimum time from an infected bite to parasitaemia is 7–8 days.) Cord blood parasites are quite frequently detected at delivery in endemic areas (Fischer, 2003; Menendez and Mayor, 2007), and detection rates are higher when sensitive PCR approaches are used (Tobian et al., 2000). Interestingly, a significant proportion of cord blood parasites appears to be acquired antenatally (Malhotra et al., 2006), although transmission at delivery also occurs.

Congenital malaria illness is more common in children of non-immune mothers. It usually presents at 2–6 weeks of age, with combinations of fever, hepatosplenomegaly, irritability, jaundice and/or anaemia. Congenital syphilis is an important differential diagnosis, and early presentations may resemble neonatal sepsis (Lesko et al., 2007; Menendez and Mayor, 2007). In the US, most cases are associated with *P. vivax* infection, reflecting the fact that many maternal infections are acquired in Latin America where *P vivax* predominates (Lesko et al., 2007). Treatment of congenital infections is with quinine for *P. falciparum* and chloroquine for *P. vivax*. Primaquine (used to eliminate liver stage *P. vivax* infections) is not required, because only blood-stage infection occurs.

13 Malaria Antibody in Neonates

In areas of high malaria transmission, not only is symptomatic congenital malaria quite uncommon, but clinical disease is also unusual in the first few months of life. A number of explanations have been proposed, including swaddling of infants (decreasing exposure to mosquitoes), parasites' inability to metabolise fetal haemoglobin effectively, and the lack of para-aminobenzoic acid (an important metabolic substrate) in breast milk (Riley et al., 2001). The relative protection from disease among infants of exposed mothers is more probably explained by the transplacental transfer of antimalarial antibodies to the fetus (Hviid and Staalsoe, 2004; Riley et al., 2001). As passively acquired IgG is catabolized and levels wane, the infant becomes more susceptible to malaria.

14 Malaria in the First Year of Life

In very young infants, asymptomatic infections may occur, and may persist for long periods, presumably contributing to early development of active immunity. Clinical studies attempting to relate placental malaria and risk of infant infections have led to somewhat confusing results. In Senegal, cord blood antibody to isolates that infect the placenta (which may not be able to sequester and survive in the baby) was associated with earlier and higher density infection in the infant, and in Cameroon, placental malaria was associated with parasitaemia until 6 months of age (Le Hesran et al., 1997). In Tanzania, there appeared to be an important interaction between gravidity and an infant's susceptibility to malaria. Whereas first time mothers are at highest risk of parasitaemia, it was the infants of multigravid women with placental infection who were at highest risk of infection in the first year of life (Mutabingwa et al., 2005).

15 Future Prospects

To prevent malaria in pregnancy, and especially to minimize more effectively its impact on young, first-time mothers and their newborns, we require improved coverage with existing interventions, such as bed nets and effective drugs for IPTp. The doses of such drugs may need to be modified for pregnant women, to ensure safe, therapeutic levels. A number of new antimalarials have entered clinical use in recent years, or will soon be introduced, and we need to evaluate carefully their safety and efficacy in pregnant women. Vaccines may have a role to play in protecting against malaria in pregnancy. Vaccines targeting pre-erythrocytic antigens (such as the RTS,S vaccine, or attenuated whole sporozoites) or conserved merozoite antigens could decrease malaria in any at-risk group, while a specific vaccine targeting the *var2csa* protein may have a specific niche in protecting pregnant women from malaria. Improvements in detecting newborns at risk of malaria in utero and after delivery may decrease the burden of morbidity and mortality in the offspring of malaria-exposed pregnant women.

Acknowledgments Stephen Rogerson is supported by the National Health and Medical Research Council of Australia and the Malaria in Pregnancy Consortium.

References

Bardaji, A., Sigauque, B., Bruni, L., Romagosa, C., Sanz, S., Mabunda, S., Mandomando, I., Aponte, J., Sevene, E., Alonso, P.L., & Menendez, C. (2008). Clinical malaria in African pregnant women. *Malar J, (7)*, 27.

Barker, D.J. (2006). Adult consequences of fetal growth restriction. *Clinical Obstetrics and Gynecology, (49)*, 270–283.

Beeson, J.G., Brown, G.V., Molyneux, M.E., Mhango, C., Dzinjalamala, F., & Rogerson, S.J. (1999). *Plasmodium falciparum* isolates from infected pregnant women and children are associated with distinct adhesive and antigenic properties. *J Infect Dis, (180)*, 464–472.

Brabin, B.J. (1983). An analysis of malaria in pregnancy in Africa. *Bull WHO, (61)*, 1005–1016.

Brabin, B.J., Agbaje, S.O., Ahmed, Y., & Briggs, N.D. (1999). A birthweight nomogram for Africa, as a malaria-control indicator. *Ann Trop Med Parasitol*, 93 (Suppl 1), S43–S57.

Brabin, B.J. & Rogerson, S.J. (2001). The epidemiology and outcomes of maternal malaria. In: P.E. Duffy, M. Fried (Eds.) *Malaria in Pregnancy Deadly Parasite, Susceptible Host*. pp. 27–52. London and New York: Taylor and Francis.

Brabin, L. (2004). The adolescent health gap in developing countries. *Ann Trop Paediatr, (24)*, 115–116.

Clark, R.L., White, T.E., S, A.C., Gaunt, I., Winstanley, P., & Ward, S.A. (2004). Developmental toxicity of artesunate and an artesunate combination in the rat and rabbit. *Birth Defects Res B Dev Reprod Toxicol, (71)*, 380–394.

Coulibaly, S.O., Gies, S., & D'Alessandro, U. (2007). Malaria burden among pregnant women living in the rural district of Boromo, Burkina Faso. *Am J Trop Med Hyg, (77)*, 56–60.

Desai, M., ter Kuile, F.O., Nosten, F., McGready, R., Asamoa, K., Brabin, B., & Newman, R.D. (2007). Epidemiology and burden of malaria in pregnancy. *Lancet Infect Dis, (7)*, 93–104.

Dondorp, A., Nosten, F., Stepniewska, K., Day, N., & White, N. (2005). Artesunate versus quinine for treatment of severe falciparum malaria: a randomised trial. *Lancet, (366)*, 717–725.

Duffy, M.F., Byrne, T.J., Elliott, S.R., Wilson, D., Rogerson, S.J., Beeson, J.G., Noviyanti, R., & Brown, G.V. (2005). Broad analysis reveals a consistent pattern of *var* gene transcription in *Plasmodium falciparum* repeatedly selected for a defined adhesion phenotype. *Mol Microbiol, (56)*, 774–788.

Duffy, M.F., Caragounis, A., Noviyanti, R., Kyriacou, H.M., Choong, E.K., Boysen, K., Healer, J., Rowe, J.A., Molyneux, M.E., Brown, G.V., & Rogerson, S.J. (2006a). Transcribed var genes associated with placental malaria in Malawian women. *Infect Immun*, 74(8), 4875–4883.

Duffy, M.F., Maier, A.G., Byrne, T.J., Marty, A.J., Elliott, S.R., O'Neill, M.T., Payne, P.D., Rogerson, S.J., Cowman, A.F., Crabb, B.S., & Brown, G.V. (2006b). *VAR2SCA* is the principal ligand for chondroitin sulfate A in two allogeneic isolates of *Plasmodium falciparum*. *Mol Biochem Parasitol, (148)*, 117–124.

Duffy, P.E. & Fried, M. (2003). Antibodies that inhibit *Plasmodium falciparum* adhesion to chondroitin sulfate A are associated with increased birth weight and the gestational age of newborns. *Infect Immun, (71)*, 6620–6623.

Fischer, P.R. (2003). Malaria and newborns. *J Trop Pediatr, (49)*, 132–134.

Francis, S.E., Malkov, V.A., Oleinikov, A.V., Rossnagle, E., Wendler, J.P., Mutabingwa, T.K., Fried, M., & Duffy, P.E. (2007). Six genes are preferentially transcribed by the circulating and sequestered forms of *Plasmodium falciparum* parasites that infect pregnant women. *Infect Immun, (75)*, 4838–4850.

Fried, M., Domingo, G.J., Gowda, C.D., Mutabingwa, T.K., & Duffy, P.E. (2006). *Plasmodium falciparum*: chondroitin sulfate A is the major receptor for adhesion of parasitized erythrocytes in the placenta. *Exp Parasitol, 113*(1), 36–42.

Fried, M. & Duffy, P.E. (1996). Adherence of *Plasmodium falciparum* to chondroitin sulfate A in the human placenta. *Science, (272)*, 1502–1504.

Fried, M., Muga, R.O., Misore, A.O., & Duffy, P.E. (1998). Malaria elicits type 1 cytokines in the human placenta: IFN-γ and TNF-α associated with pregnancy outcomes. *J Immunol, (160)*, 2523–2530.

Garner, P., & Gulmezoglu, A.M. (2006). Drugs for preventing malaria in pregnant women. *Cochrane Database Syst Rev,* CD000169.

Genton, B., D'Acremont, V., Rare, L., Baea, K., Reeder, J.C., Alpers, M.P., & Mueller, I.M. (2008). *Plasmodium vivax* and mixed infections are associated with severe malaria in children: a prospective cohort study from Papua New Guinea. *PLoS Med, (5)*, e127.

Green, M.D., van Eijk, A.M., van Ter Kuile, F.O., Ayisi, J.G., Parise, M.E., Kager, P.A., Nahlen, B.L., Steketee, R., & Nettey, H. (2007). Pharmacokinetics of sulfadoxine-pyrimethamine in HIV-infected and uninfected pregnant women in Western Kenya. *J Infect Dis, (196)*, 1403–1408.

Guyatt, H.L. & Snow, R.W. (2001). The epidemiology and burden of *Plasmodium falciparum*-related anemia among pregnant women in sub-Saharan Africa. *Am J Trop Med Hyg,* 64 (Suppl), 36–44.

Guyatt, H.L. & Snow, R.W. (2004). Impact of malaria during pregnancy on low birth weight in sub-Saharan Africa. *Clin Microbiol Rev, (17)*, 760–769, table of contents.

Haase, R.N., Megnekou, R., Lundquist, M., Ofori, M.F., Hviid, L., & Staalsoe, T. (2006). *Plasmodium falciparum* parasites expressing pregnancy-specific variant surface antigens adhere strongly to the choriocarcinoma cell line BeWo. *Infect Immun, (74)*, 3035–3038.

Hart, C.A. (1988). Pregnancy and host resistance. *Baillieres Clin Immunol Allergy, (2)*, 735–757.

Hviid, L. (2004). The immuno-epidemiology of pregnancy-associated *Plasmodium falciparum* malaria: a variant surface antigen-specific perspective. *Parasite Immunol, (26)*, 477–486.

Hviid, L. & Staalsoe, T. (2004). Malaria immunity in infants: a special case of a general phenomenon? *Trends Parasitol, (20)*, 66–72.

Le Hesran, J.Y., Cot, M., Personne, P., Fievet, N., Dubois, B., Beyeme, M., Boudin, C., & Deloron, P. (1997). Maternal placental infection with *Plasmodium falciparum* and malaria morbidity during the first 2 years of life. *Am J Epidemiol, (146)*, 826–831.

Lesko, C.R., Arguin, P.M., & Newman, R.D. (2007). Congenital malaria in the United States: a review of cases from 1966 to 2005. *Arch Pediatr Adolesc Med, (161)*, 1062–1067.

Lindsay, S., Ansell, J., Selman, C., Cox, V., Hamilton, K., & Walraven, G. (2000). Effect of pregnancy on exposure to malaria mosquitoes. *Lancet, (355)*, 1972.

Lucchi, N.W., Koopman, R., Peterson, D.S., & Moore, J.M. (2006). *Plasmodium falciparum*-infected red blood cells selected for binding to cultured syncytiotrophoblast bind to chondroitin sulfate A and induce tyrosine phosphorylation in the syncytiotrophoblast. *Placenta, (27)*, 384–394.

Malhotra, I., Mungai, P., Muchiri, E., Kwiek, J.J., Meshnick, S.R., & King, C.L. (2006). Umbilical cord-blood infections with *Plasmodium falciparum* malaria are acquired antenatally in kenya. *J Infect Dis, (194)*, 176–183.

Marsh, K. & Kinyanjui, S. (2006). Immune effector mechanisms in malaria. *Parasite Immunol, (28)*, 51–60.

Maubert, B., Fievet, N., Tami, G., Cot, M., Boudin, C., & Deloron, P. (1999). Development of antibodies against chondroitin sulfate A-adherent *Plasmodium falciparum* in pregnant women. *Infect Immun, (67)*, 5367–5371.

Menendez, C. & Mayor, A. (2007). Congenital malaria: the least known consequence of malaria in pregnancy. *Semin Fetal Neonatal Med, (12)*, 207–213.

Menendez, C., Ordi, J., Ismail, M.R., Ventura, P.J., Aponte, J.J., Kahigwa, E., Font, F., & Alonso, P.L. (2000). The impact of placental malaria on gestational age and birth weight. *J Infect Dis,* (181), 1740–1745.

Moormann, A.M., Sullivan, A.D., Rochford, R.A., Chensue, S.W., Bock, P.J., Nyirenda, T., & Meshnick, S.R. (1999). Malaria and pregnancy: placental cytokine expression and its relationship to intrauterine growth retardation. *J Infect Dis,* (180), 1987–1993.

Mutabingwa, T.K., Bolla, M.C., Li, J.L., Domingo, G.J., Li, X., Fried, M., & Duffy, P.E. (2005). Maternal malaria and gravidity interact to modify infant susceptibility to malaria. *PLoS Med,* (2), e407.

Nosten, F., McGready, R., Simpson, J.A., Thwai, K.L., Balkan, S., Cho, T., Hkirijaroen, L., Looareesuwan, S., & White, N.J. (1999). Effects of *Plasmodium vivax* malaria in pregnancy. *Lancet,* (354), 546–549.

Parise, M.E., Ayisi, J.G., Nahlen, B.L., Schultz, L.J., Roberts, J.M., Misore, A., Muga, R., Oloo, A.J., & Steketee, R.W. (1998). Efficacy of sufadoxine-pyrimethamine for prevention of placental malaria in an area of Kenya with a high prevalence of malaria and human immunodeficiency virus infection. *Am J Trop Med Hyg,* (59), 813–822.

Phillips-Howard, P.A. & Wood, D. (1996). The safety of antimalarial drugs in pregnancy. *Drug Saf,* (14), 131–145.

Poespoprodjo, J.R., Fobia, W., Kenangalem, E., Lampah, D.A., Warikar, N., Seal, A., McGready, R., Sugiarto, P., Tjitra, E., Anstey, N.M., & Price, R.N. (2008). Adverse pregnancy outcomes in an area where multidrug-resistant *plasmodium vivax* and *Plasmodium falciparum* infections are endemic. *Clin Infect Dis,* (46), 1374–1381.

Riley, E.M., Wagner, G.E., Akanmori, B.D., & Koram, K.A. (2001). Do maternally acquired antibodies protect infants from malaria infection? *Parasite Immunol,* (23), 51–59.

Rogerson, S.J., Brown, H.C., Pollina, E., Abrams, E.T., Tadesse, E., Lema, V.M., & Molyneux, M.E. (2003a). Placental tumor necrosis factor alpha but not gamma interferon is associated with placental malaria and low birth weight in Malawian women. *Infect Immun,* (71), 267–270.

Rogerson, S.J., Chaiyaroj, S.C., Ng, K., Reeder, J.C., & Brown, G.V. (1995). Chondroitin sulfate A is a cell surface receptor for *Plasmodium falciparum*-infected erythrocytes. *J Exp Med,* (182), 15–20.

Rogerson, S.J., Chaluluka, E., Kanjala, M., Mkundika, P., Mhango, C.G., & Molyneux, M.E. (2000). Intermittent sulphadoxine-pyrimethamine in pregnancy: effectiveness against malaria morbidity in Blantyre, Malawi 1997–1999. *Trans R Soc Trop Med Hyg,* (94), 549–553.

Rogerson, S.J., Hviid, L., Duffy, P.E., Leke, R.F., & Taylor, D.W. (2007). Malaria in pregnancy: pathogenesis and immunity. *Lancet Infect Dis,* (7), 105–117.

Rogerson, S.J. & Menendez, C. (2006). Treatment and prevention of malaria in pregnancy: opportunities and challenges. *Expert Rev Anti Infect Ther,* (4), 687–702.

Rogerson, S.J., Pollina, E., Getachew, A., Tadesse, E., Lema, V.M., & Molyneux, M.E. (2003b). Placental monocyte infiltrates in response to *Plasmodium falciparum* infection and their association with adverse pregnancy outcomes. *Am J Trop Med Hyg,* (68), 115–119.

Salanti, A., Dahlback, M., Turner, L., Nielsen, M.A., Barfod, L., Magistrado, P., Jensen, A.T., Lavstsen, T., Ofori, M.F., Marsh, K., Hviid, L., & Theander, T.G. (2004). Evidence for the Involvement of *VAR2SCA* in pregnancy-associated Malaria. *J Exp Med,* (200), 1197–1203.

Salanti, A., Staalsoe, T., Lavstsen, T., Jensen, A.T., Sowa, M.P., Arnot, D.E., Hviid, L., & Theander, T.G. (2003). Selective upregulation of a single distinctly structured *var* gene in chondroitin sulphate A-adhering *Plasmodium falciparum* involved in pregnancy-associated malaria. *Mol Microbiol,* (49), 179–191.

Saute, F., Menendez, C., Mayor, A., Aponte, J., Gomez-Olive, X., Dgedge, M., & Alonso, P. (2002). Malaria in pregnancy in rural Mozambique: the role of parity, submicroscopic and multiple *Plasmodium falciparum* infections. *Trop Med Int Health,* (7), 19–28.

Singh, B., Kim Sung, L., Matusop, A., Radhakrishnan, A., Shamsul, S.S., Cox-Singh, J., Thomas, A., & Conway, D.J. (2004). A large focus of naturally acquired Plasmodium knowlesi infections in human beings. *Lancet,* (363), 1017–1024.

Singh, N., Shukla, M.M., & Sharma, V.P. (1999). Epidemiology of malaria in pregnancy in central India. *Bull WHO,* (77), 567–572.

Snow, R.W., Guerra, C.A., Noor, A.M., Myint, H.Y., & Hay, S.I. (2005). The global distribution of clinical episodes of *Plasmodium falciparum* malaria. *Nature,* (434), 214–217.

Staalsoe, T., Shulman, C.E., Bulmer, J.N., Kawuondo, K., Marsh, K., & Hviid, L. (2004). Variant surface antigen-specific IgG and protection against clinical consequences of pregnancy-associated *Plasmodium falciparum* malaria. *Lancet,* (363), 283–289.

Steketee, R.W., Nahlen, B.L., Parise, M.E., & Menendez, C. (2001). The burden of malaria in pregnancy in malaria-endemic areas. *Am J Trop Med Hyg,* (64), 28–35.

Suguitan, A.L., Jr., Leke, R.G., Fouda, G., Zhou, A., Thuita, L., Metenou, S., Fogako, J., Megnekou, R., & Taylor, D.W. (2003). Changes in the levels of chemokines and cytokines in the placentas of women with *Plasmodium falciparum* malaria. *J Infect Dis,* (188), 1074–1082.

Tagbor, H., Bruce, J., Browne, E., Greenwood, B., & Chandramohan, D. (2008). Malaria in pregnancy in an area of stable and intense transmission: is it asymptomatic? *Trop Med Int Health,* 13(8), 1016-1021.

ter Kuile, F.O., van Eijk, A.M., & Filler, S.J. (2007). Effect of sulfadoxine-pyrimethamine resistance on the efficacy of intermittent preventive therapy for malaria control during pregnancy: a systematic review. *Jama,* (297), 2603–2616.

Tjitra, E., Anstey, N.M., Sugiarto, P., Warikar, N., Kenangalem, E., Karyana, M., Lampah, D.A., & Price, R.N. (2008). Multidrug-resistant *Plasmodium vivax* associated with severe and fatal malaria: A prospective study in Papua, Indonesia. *PLoS Med,* (5), e128.

Tobian, A.A.R., Mehlotra, R.K., Malhotra, I., Wamachi, A., Mungai, P., Koech, D., Ouma, J., Zimmerman, P., & King, C.L. (2000). Frequent umbilical cord-blood and maternal-blood infections with *Plasmodium falciparum, P. malariae* and *P. ovale* in Kenya. *J Infect Dis,* (182), 558–563.

Tuikue Ndam, N., Bischoff, E., Proux, C., Lavstsen, T., Salanti, A., Guitard, J., Nielsen, M.A., Coppee, J.Y., Gaye, A., Theander, T., David, P.H., & Deloron, P. (2008). *Plasmodium falciparum* transcriptome analysis reveals pregnancy malaria associated gene expression. *PLoS ONE,* (3), e1855.

Tuikue Ndam, N.G., Salanti, A., Bertin, G., Dahlback, M., Fievet, N., Turner, L., Gaye, A., Theander, T., & Deloron, P. (2005). High level of *var2sca* transcription by *Plasmodium falciparum* isolated from the placenta. *J Infect Dis,* (192), 331–335.

Viebig, N.K., Gamain, B., Scheidig, C., Lepolard, C., Przyborski, J., Lanzer, M., Gysin, J., & Scherf, A. (2005). A single member of the *Plasmodium falciparum* var multigene family determines cytoadhesion to the placental receptor chondroitin sulphate A. *EMBO Rep,* (6), 775–781.

Vleugels, M.P.H., Brabin, B., Eling, W.M.C., & de Graaf, R. (1989). Cortisol and *Plasmodium falciparum* infection in pregnant women in Kenya. *Trans R Soc Trop Med Hyg,* (83), 173–177.

Walter, P.R., Garin, Y., & Blot, P. (1982). Placental pathologic changes in malaria. A histologic and ultrastructural study. *Am J Pathol,* (109), 330–342.

Ward, S.A., Sevene, E.J., Hastings, I.M., Nosten, F., & McGready, R. (2007). Antimalarial drugs and pregnancy: safety, pharmacokinetics, and pharmacovigilance. *Lancet Infect Dis,* (7), 136–144.

Yosaatmadja, F., Andrews, K.T., Duffy, M.F., Brown, G.V., Beeson, J.G., & Rogerson, S.J. (2008). Characterization of *VAR2SCA*-deficient *Plasmodium falciparum*-infected erythrocytes selected for adhesion to the BeWo placental cell line. *Malar J,* (7), 51.

Adenovirus Infection in the Immunocompromised Host

Marc Tebruegge and Nigel Curtis

1 Classification, Structure, and Molecular Pathogenesis

There are 51 known serotypes of human adenovirus, which are divided into subgroups (or species) A to F (Table 1). An additional serotype (52), which potentially constitutes a new species (G), has recently been reported after genomic sequencing and phylogenetic analysis of an isolate in the U.S. (Jones et al., 2007).

Adenovirus is an icosahedral, non-enveloped virus consisting of a nucleocapsid and a DNA genome (Medina-Kauwe, 2003). The double-stranded DNA is 30–38 kilobases in size and encodes more than 30 structural and non-structural proteins. The capsid is primarily composed of hexon protein (Fig. 1). Each vertex of the icosahedron is capped by a penton base, comprising a pentameric ring of proteins. From the centre of this base, arises the trimeric penton fiber. This fiber is responsible for the initial attachment of adenoviruses to the host cell and has been shown to bind to cocksackie-adenovirus receptor (CAR) cell surface protein with high affinity (Bergelson et al., 1997). This initial binding is followed by an interaction between penton proteins and integrin co-receptors ($\alpha_v\beta_3$ and $\alpha_v\beta_5$), which triggers integrin-mediated endocytosis (Nemerow et al., 1994; Wickham et al., 1993).

After entering the cell, the virus evades lysosomal degradation by penetrating the endosomal membrane and thereby escapes into the cysol (Seth, 1994). The capsid then translocates towards the nucleus and docks onto nuclear pore complexes, which is followed by extrusion of viral DNA and protein VII into the nucleus (Greber et al., 1993; 1997). After reaching the nucleus, viral genes are expressed and new virus particles are generated.

N. Curtis (✉)

Department of Paediatrics, The University of Melbourne; Infectious Diseases Unit, Department of General Medicine; Microbiology & Infectious Diseases Research Group, Murdoch Children's Research Institute: Royal Children's Hospital Melbourne, Parkville, VIC 3052, Australia
e-mail: nigel.curtis@rch.org.au

A. Finn et al. (eds.), *Hot Topics in Infection and Immunity in Children VI*, Advances in Experimental Medicine and Biology 659, DOI 10.1007/978-1-4419-0981-7_13, © Springer Science+Business Media, LLC 2010

Table 1 Classification of adenoviruses and common sites of infection

Subgroup	Serotypes	Sites
A	12, 18, 31	Gastrointestinal tract
B	3, 7, 11, 14, 16, 21, 34, 35, 50	
C	1, 2, 5, 6	Respiratory tract
D	8, 9, 10, 13, 15, 17, 19, 20, 22–30, 32, 33, 36–39, 42–49, 51	Ophthalmic, gastrointestinal tract
E	4	Respiratory tract
F	40, 41	Gastrointestinal tract

Fig. 1 The structure of adenovirus and binding to the host cell

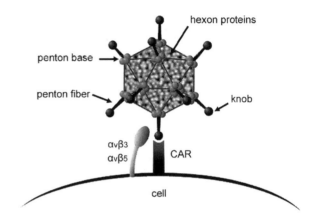

2 Terminology and Definitions Related to Adenovirus Infection and Disease

In discussing adenovirus infection and disease it is important to clarify definitions, particularly as the existing literature is complicated by the use of imprecise terminology. The majority of authors have adhered to the definitions originally proposed by Flomenberg et al. in 1994 (Table 2) (Flomenberg et al., 1994), although frequently in a heavily modified form. These definitions, however, have two major limitations. Firstly, they rely on the use of biopsies which are invasive procedures that are not frequently performed in a paediatric setting. Secondly, the criteria do not include results of polymerase chain reaction (PCR) tests, which have subsequently become part of standard practice. We therefore prefer the following definitions, which will be used throughout this review (Table 3): (i) *adenovirus infection*: detection of adenovirus by immunofluorescence, culture, histology, or PCR from any site (ii) *asymptomatic adenovirus infection*: detection of adenovirus infection from any site in the absence of clinical symptoms (iii) *adenovirus disease* (i.e., 'symptomatic adenovirus infection'): detection of adenovirus from any site with anatomically corresponding, compatible symptoms in the absence of any other identifiable cause

Table 2 Criteria for adenovirus disease proposed by Flomenberg et al., 1994

Definite disease	Presence of typical adenovirus nuclear inclusions on routine histopathology *or* positive culture from tissue (excluding gastrointestinal tract)
Probable disease	Two or more positive cultures from body sites other than the gastro-intestinal tract *plus* compatible symptoms without other identifiable causes

Table 3 Proposed new adenovirus definitions (as used in this review)

Adenovirus infection	Detection of adenovirus by immunofluorescence, culture, histology or PCR from any site
Asymptomatic adenovirus infection	Detection of adenovirus from any site in the absence of clinical symptoms
Adenovirus disease	Detection of adenovirus from any site with anatomically corresponding, compatible symptoms in the absence of any other identifiable cause
Disseminated adenovirus disease	Adenovirus disease affecting two or more organs or systems

and (iv) *disseminated adenovirus disease*: adenovirus disease affecting two or more organs or systems.

3 Epidemiology, Clinical Features, and Natural History in the General Population

Adenovirus can be transmitted from person to person via respiratory droplets, as well as by the conjunctival and the faecal-oral route. The incubation period is generally short, ranging from 2 to 14 days. The majority of adenovirus infections occur in the first 5 years of life, with a peak incidence during the first 2 years (Hong et al., 2001; Pacini et al., 1987).

Adenovirus infections are very common in children. Epidemiological data suggest that around 5–15% of acute upper respiratory tract infections and around 5% of lower respiratory tract infections in childhood are caused by this virus (Edwards et al., 1985; Gardner, 1968; Hong et al., 2001; Kim et al., 2000).

In immunocompetent individuals, adenovirus infections are generally mild and self-limiting, manifesting primarily as upper respiratory tract infection (tonsillitis, pharyngitis, otitis media) and bronchiolitis/bronchitis, gastroenteritis, or keratoconjunctivitis (Edwards et al., 1985; Gardner, 1968; Pacini et al., 1987). Fatal outcome in immunocompetent individuals is relatively rare and has primarily been reported in the context of pneumonitis (Hong et al., 2001). However, recent reports from the U.S. have described the emergence of a new, potentially more aggressive adenovirus serotype 14 variant (Ad14). Strikingly, a significant proportion (17%) of Ad14 infected patients were reported to require intensive care support, and the overall fatality rate was as high as 5% (CDC, 2007; Louie et al., 2008).

Different serotypes have some degree of tropism for different types of tissue and are therefore associated with different clinical manifestation. The majority of lower respiratory tract infections in children caused by adenovirus are due to subgroup B serotypes 3, 7, and 21 (Hong et al., 2001). Subgroup C serotypes 1, 2, and 5 are the serotypes that are most commonly associated with upper respiratory tract infections in childhood (Gardner, 1968; Pacini et al., 1987). Serotypes from subgroups D and F are implicated in gastrointestinal infection, while epidemic keratoconjunctivitis is predominantly caused by subgroup D serotypes.

Primary adenovirus infection results in the production of neutralising antibodies and is thought to confer lifelong immunity against the causal serotype (Edwards et al., 1985). However, despite the presence of neutralising antibodies, some patients shed adenovirus in the stool for several months following primary infection. Other patients develop longstanding, potentially lifelong, asymptomatic infection with persistence of adenovirus in lympho-epithelial tissues (Adrian et al., 1988; Fox et al., 1969; Garnett et al., 2002). The latter has direct implications for patients undergoing transplantation or immunosuppression, as this may result in endogenous reactivation of latent adenovirus infection.

4 Epidemiology, Clinical Features, and Natural History in the Immununocompromised

Since the early 1990s, adenovirus has become increasingly recognised as an important pathogen in immunocompromised hosts in general and, in particular, in transplant recipients (Flomenberg et al., 1994; Hierholzer, 1992; Walls et al., 2003). Cumulative data suggest that adenovirus infection occurs in up to 40% of paediatric human stem cell transplant (HSCT) recipients (Kampmann et al., 2005; Lion et al., 2003; Symeonidis et al., 2007; Walls et al., 2003), around 10% of adult HSCT recipients (Baldwin et al., 2001; Bruno et al., 2003; Flomenberg et al., 1994; Kalpoe et al., 2007; Runde et al., 2001) and 5–10% of solid organ transplant recipients (de Mezerville et al., 2006; Hoffman, 2006; Humar et al., 2005; Ison, 2006). In these patient groups, adenovirus infection commonly produces more severe disease than in immunocompetent individuals. In this setting, the most common clinical manifestations are haemorrhagic cystitis and haemorrhagic enteritis, followed by pneumonitis, hepatitis, nephritis, encephalitis, and multi-organ failure (Baldwin et al., 2001; Chakrabarti et al., 2002; de Mezerville et al., 2006; Feuchtinger et al., 2007; Ison, 2006; La Rosa et al., 2001).

In solid organ transplant recipients, there is a tendency for adenovirus infection and disease to primarily involve the transplanted organ (Kojaoghlanian et al., 2003). Therefore, hepatitis is a common manifestation in liver transplant recipients, enteritis in small bowel transplants, and heamorrhagic cystitis in renal transplants. This association strongly suggests that reactivation of latent virus in the transplanted organ plays an important role. While any organ or system may be affected by adenovirus in a HSCT setting, disease most commonly occurs in the gut and bladder.

The majority of adenovirus disease in the transplant setting occurs during the first 100 days post-transplant (Bordigoni et al., 2001; Flomenberg et al., 1994; Hale et al., 1999; Howard et al., 1999; La Rosa et al., 2001; Runde et al., 2001). Overall, disease tends to occur earlier in children compared to adults. Nevertheless, severe adenovirus disease, with fatal outcome in some cases, later than 1 year following transplantation has been reported in several patients (Erard et al., 2007; Flomenberg et al., 1994; Ljungman et al., 2003).

In the transplant setting, progression of disease with subsequent involvement of other organs and/or systems is common. Notably, data from historical studies suggest that without antiviral treatment as many as half of the cases with initially localised adenovirus disease develop disseminated disease (Flomenberg et al., 1994). HSCT patients are more likely to develop multi-organ disease compared to solid organ transplant recipients (de Mezerville et al., 2006).

Severe adenovirus disease has also been described in other immunocompromised patient groups, including patients with HIV-infection (Adeyemia et al., 2008; Nebbia et al., 2005), patients with malignancies receiving chemotherapy, (Fianchi et al., 2003; Hough et al., 2005; Steiner et al., 2008), and in primary immunodeficiencies (Dagan et al., 1984; Wigger and Blanc, 1966), although these account for a relatively small proportion of the total number of cases with severe adenovirus disease that is reported in the literature.

5 Risk Factors for Adenovirus Infection in the Transplant Setting

In recent years, there has been mounting evidence that in bone marrow transplant recipients T cell depletion constitutes the single most important risk factor for adenovirus infection. Several reports have shown that the use of T cell depleted grafts (Chakrabarti et al., 2002; Kampmann et al., 2005; Lion et al., 2003), as well as drugs in the conditioning regimen that reduce the T cell population in the recipient, such as alemtuzumab (campath®/CD52-directed cytolytic antibody) and anti-thymocyte globulin (ATG) (Chakrabarti et al., 2002; Sivaprakasam et al., 2007), are strongly associated with the development of adenovirus disease.

Other risk factors include young age (children compared to adults) (Baldwin et al., 2001; Bruno et al., 2003), allogenic HSCT (compared to autologous) (Howard et al., 1999; Muller et al., 2005), and the use of a matched unrelated donor (compared to matched sibling donors) (Baldwin et al., 2001; Kampmann et al., 2005; Lion et al., 2003).

There are conflicting data in adenovirus infection, there is conflicting data regarding the role of graft-versus-host-disease (GvHD). One study in paediatric HSCT patients found that as many as 26% of patients with acute GvHD developed adenovirus viraemia (Kampmann et al., 2005). However, while several studies have reported an increased risk of adenovirus infection and disease in patients with significant GvHD (Bruno et al., 2003; Flomenberg et al., 1994; Kampmann et al., 2005; Shields et al., 1985), others have not observed this association (Lion et al., 2003).

In solid organ transplantation, analogous to the observations in HSCT, adenovirus infection is also more common in transplant recipients in the paediatric age group compared to adults (Koneru et al., 1987; McGrath et al., 1998; Michaels et al., 1992). In solid organ transplant, both the use of muromonab (OKT3®/anti-CD3 monoclonal antibody) and anti-thymocyte globulin (ATG), given as an intervention for steroid-resistant graft rejection, have been identified as risk factors for invasive adenovirus disease (Hoffman, 2006; Koneru et al., 1987; Michaels et al., 1992).

6 Diagnostic Methods in Adenovirus Infection

A variety of methods facilitate the identification of adenovirus infection. These include culture of throat swabs, stool, urine, and biopsy material in appropriate tissue culture medium (e.g., human embryonic kidney, HEp-2, HeLa or MRC-5 cells), which allows subsequent serotyping (Mahafzah and Landry, 1989). However, identification of adenovirus in traditional culture systems can take up to 5–10 days, depending on the cell types used, thus leading to a significant delay in diagnosis (Mahafzah and Landry, 1989). Shell vial assays have been shown to expedite detection while maintaining high levels of sensitivity (97–98% after 48 h) (Espy et al., 1987).

Alternatively, samples can be directly stained in immunofluorescence assays using monoclonal antibodies to detect the presence of adenovirus antigen. This method of is primarily used for samples from the respiratory tract (e.g., nasopharyngeal aspirate, bronchoalveolar lavage fluid) but can also be applied to conjunctival scrapings. However, while these assays allow rapid detection of adenovirus infection, and generally have high specificity, their sensitivity (around 70–85%) is inferior to culture-based methods (El-Sayed Zaki and Abd-El Fatah, 2008; Freymuth et al., 2006).

Serology can also be used, although a positive result at a single point in time has meaning, given that adenovirus infections in early childhood are very common (ie a positive result may merely reflect past exposure). However, paired serum samples showing a significant rise in antibody titre may provide useful information.

PCR has been shown to have both high sensitivity and specificity for the diagnosis of adenovirus infection (Echavarria et al., 1998; El-Sayed Zaki and Abd-El Fatah, 2008; Freymuth et al., 2006). Amplification of adenoviral DNA is generally achieved by using primers that bind to highly conserved regions within the hexon gene (Echavarria et al., 1998). The causative serotype can be determined using subsequent sequence analysis (Wong et al., 2008). This method is ideally suited for the analysis of samples from normally sterile sites such as cerebrospinal fluid and blood. Results from upper respiratory tract and stool samples are difficult to interpret, as a positive PCR may simply represent viral shedding, rather than correlate with adenoviral disease (see criteria defining infection/disease above) (Garnett et al., 2002). Interestingly, a recent study analysing bronchoalveolar lavage samples in individuals without evidence of adenovirus disease has shown that positive PCR results can occur in the absence of clinical symptoms, suggesting that adenovirus may also persist in the lower respiratory tract (Leung et al., 2005).

In recent years many research groups, as well as transplant centres, have started to use real-time PCR for the detection and monitoring of adenovirus on a routine basis. This method allows quantification of adenoviral DNA (i.e., viral load (VL)) and is particularly useful for assessing the kinetics of adenovirus viraemia. Several reports have shown that serial monitoring of VL can provide crucial information and augment management decisions, particularly in relation to antiviral therapy. Notably, several reports document that a 10-fold rise in copy number as well as a VL of $> 10^6$ copies/ml are associated with progression of adenovirus infection and poor outcome (Claas et al., 2005; Lankester et al., 2002; Lion et al., 2003; Neofytos et al., 2007; Schilham et al., 2002; Seidemann et al., 2004). Other studies have shown that VL can be used to assess treatment response by demonstrating a correlation between VL and clinical recovery (Lankester et al., 2004; Leruez-Ville et al., 2004; Neofytos et al., 2007). Notably, in a study by Leruez-Ville et al., a greater than 10-fold decrease in VL 7–10 days after the first dose of antiviral therapy was associated with a favourable clinical course (Leruez-Ville et al., 2004). In contrast, all patients with a fatal outcome in this study failed to show a significant reduction in VL after treatment was initiated.

7 Prognosis of Adenovirus Infection in the Immunocompromised

Overall, the mortality in HSCT patients with adenovirus infection is high, ranging from 12 to 60% in different studies (Flomenberg et al., 1994; Hierholzer, 1992; Howard et al., 1999; Sivaprakasam et al., 2007; Symeonidis et al., 2007). Poor prognostic indicators include disseminated disease, hepatitis, and pneumonia, which have been found to be associated with mortality rates of approximately 50, 70, and 75% respectively (Bruno et al., 2003; Hough et al., 2005; Ison, 2006; La Rosa et al., 2001). In general, adenovirus disease in children undergoing HSCT carries a worse prognosis than in the adult transplant population. This may reflect a higher rate of *de novo* infections in children, compared to a larger proportion of reactivated latent infections in adult patients.

In lung transplant recipients, adenovirus infection is associated with an equally poor prognosis (Matar et al., 1999; Palmer et al., 1998). In this patient group, respiratory tract infections due to adenovirus have a tendency to progress rapidly, both radiologically and clinically. The currently available data, based on relatively small patient numbers, suggest that the adenovirus-associated mortality in these patients may be as high as 33–80% (Matar et al., 1999; Palmer et al., 1998). In addition, there is some evidence that a large proportion of survivors develop chronic obliterative bronchiolitis, in some cases requiring re-transplantation. (Doan et al., 2007; Palmer et al., 1998).

In cardiac transplant recipients, adenovirus infection has been described to be one of the most common causes of late or chronic graft rejection based on endomyocardial biopsies (Schowengerdt et al., 1996; Shirali et al., 2001). One study reported that detection of adenoviral DNA in PCR was linked to coronary vasculopathy and strongly associated with reduced graft survival (Shirali et al., 2001).

There are only limited data related to adenovirus infection in orthotopic liver transplant (OLT) recipients. Although there have been several case reports, there are only a few cohort studies published to date that allow a more comprehensive view (Cames et al., 1992; Koneru et al., 1987; McGrath et al., 1998; Michaels et al., 1992). In the only cohort study in adults, three of 11 (27%) OLT recipients who acquired adenovirus infection in the post-transplant period died of adenovirus-related complications, primarily comprising hepatitis and pneumonitis (McGrath et al., 1998). In the paediatric OLT setting the adenovirus-related mortality ranges between 9 and 18%, with fulminant hepatitis – frequently caused by serotype 5 – constituting the most common cause of death (Cames et al., 1992; Koneru et al., 1987; Michaels et al., 1992).

Adenovirus infection is also a relatively common complication in intestinal transplant recipients (Adeyi et al., 2008; Berho et al., 1998; Parizhskaya et al., 2001; Pinchoff et al., 2003; Ziring et al., 2005). Adenovirus was the second most common infectious cause of enteritis in this patient group following rotavirus in one study (Ziring et al., 2005). Many of the adenovirus-related episodes in this study were initially misinterpreted as acute graft rejection, a finding reflected in other reports (Adeyi et al., 2008; Parizhskaya et al., 2001). One case in this series was diagnosed only after removal of the allograft for what was thought to be refractory acute rejection. Strikingly, another study, which included 14 small bowel transplant recipients that underwent active viral surveillance post-transplantation, found that all patients had at least one positive intestinal adenoviral culture (serotypes 1, 5, and 31) (Pinchoff et al., 2003). In eight of these patients adenovirus enteritis was confirmed on histology. Four of these 14 patients (29%) died. In two cases the death was directly related to adenovirus; in the remaining two the role of adenovirus was less certain.

The literature related to adenovirus infection following renal transplants consists mainly of case reports and small case series, which prevents firm conclusions about the overall mortality in this setting (Ardehali et al., 2001; Keswani and Moudgil, 2007; Myerowitz et al., 1975; Rosario et al., 2006). However, a review of the adult literature concluded that spontaneous resolution of haemorrhagic cystitis in this patient group was common (Hofland et al., 2004). Nevertheless, several authors have reported cases with a fatal outcome (Ardehali et al., 2001; Myerowitz et al., 1975; Rosario et al., 2006), suggesting that these patients may also benefit from active intervention.

8 Control of Adenovirus Infection and Immune Recovery

Over the last decade there has been mounting evidence from both animal models and human studies, that control and elimination of adenovirus infection is primarily related to immune recovery. The principal role of antiviral drugs appears to be the reduction of viral replication until this has occurred.

A fascinating study in a severe combined immunodeficiency (SCID) murine model showed that antiviral therapy with cidofovir leads to a significant delay in disease progression (Lenaerts et al., 2005). However, all animals in this study ultimately died, leading the authors to conclude that some degree of adaptive immune response is required to eliminate the virus.

Chakrabarti et al. were the first to show that adenovirus disease is highly significantly associated with slow lymphocyte count recovery following HSCT (Chakrabarti et al., 2002). Similar observations were subsequently reported by other groups (Heemskerk et al., 2005; Kalpoe et al., 2007; Kampmann et al., 2005). One study reported the risk of death in transplant patients with adenovirus infection and an absolute lymphocyte count of less than 300/µl to be as high as 47% (Chakrabarti et al., 2002).

Interestingly, van Tol et al. reported that slow T cell recovery in HSCT patients, based on the counts of CD3+ T cells or CD4+ and CD8+ T-cell subsets, correlated significantly with the incidence of adenovirus disease (van Tol et al., 2005). In addition, slow lymphocyte recovery was associated with fatal outcome. Furthermore, failure to engraft was associated with a 10-fold increased risk (95% CI 1.0–97.6) of adenovirus disease and a 35-fold risk (95% CI 4.0–309.1) of adenovirus-related death in this report. Later work by the same group – also in HSCT patients – suggests that the ability to generate adenovirus-specific CD4+ T cells plays a crucial role in the control of adenovirus infection (Heemskerk et al., 2005).

Finally, multiple studies have shown that reduction of immunosuppression in transplant patients has a clear beneficial effect with regard to adenovirus infection (Chakrabarti et al., 2002; Kampmann et al., 2005; Sivaprakasam et al., 2007; van Tol et al., 2005; Ziring et al., 2005). Conversely, failure to reduce immunosuppression, commonly in patients in whom this was not possible due to severe GvHD, was associated with poor outcome and high levels of mortality.

9 Indications for Treatment of Adenovirus Infection

As outlined above, there are many variables that have to be taken into account when considering treatment of adenovirus infection, such as age (higher mortality in children), state of and prospect for immune recovery, site of the disease (higher mortality in pneumonia, hepatitis and multiorgan-disease) and the type of transplant (higher mortality in HSCT and lung transplants).

Notably, data from some of the earlier studies suggested that isolation of adenovirus from two or more sites is the *only* reliable predictive marker for severe adenovirus disease and poor outcome (Baldwin et al., 2001; Flomenberg et al., 1994; Howard et al., 1999). Furthermore, a number of studies demonstrated that adenovirus disease develops in 60–100% of patients in whom adenovirus can be isolated from multiple sites (Chakrabarti et al., 2002; Flomenberg et al., 1994; Hoffman et al., 2001; Howard et al., 1999). Unfortunately, this paradigm continues to be used in some centres as the main, or even only, indication to initiate treatment, despite the fact that data from more recent studies indicate that the issue is more complex.

Most experts agree that detection of adenovirus from nasopharyngeal, urine or stool samples by immunofluorescence, culture or PCR in an asymptomatic patient does not necessarily warrant treatment, as this may simply represent viral shedding. Few would disagree with the notion of initiating treatment in immunocompromised patients with symptomatic disease, in whom adenovirus is identified from the corresponding site in the absence of an alternative explanation. However, currently there is no consensus regarding the treatment of asymptomatic patients who are found to have adenovirus viraemia.

Walls et al. suggested that some patients with adenovirus viraemia may not require antiviral therapy, as a significant proportion (64%) of paediatric HSCT recipients cleared the virus without specific therapy in their study (Walls et al., 2005). Nevertheless, a large number of reports have shown a close correlation between adenovirus viraemia and fatal outcome (Chakrabarti et al., 2002; Echavarria et al., 2001; Lankester et al., 2004; Lion et al., 2003; Schilham et al., 2002). Notably, in one study, 73% of paediatric HSCT patients in whom adenovirus was detected in blood by PCR had a fatal outcome directly related to adenovirus (Lion et al., 2003). Similarly, another study reported that four of six initially asymptomatic viraemic patients developed symptomatic disease within 1–2 weeks, finally resulting in the death of three patients (Seidemann et al., 2004). In the study by Leruez-Ville et al. the mortality in blood PCR-positive patients was considerably lower (25%), although it appears likely that this was influenced by the initiation of antiviral therapy in many cases (Leruez-Ville et al., 2004).

There is some evidence that in adult HSCT patients adenovirus viraemia is not as closely related to poor prognosis (Kalpoe et al., 2007). Similarly, in solid organ transplant recipients, viraemia is not as closely linked to adverse outcome. Notably, in one report, which included 19 adult solid organ transplant recipients, all patients cleared adenovirus viraemia without specific treatment (Humar et al., 2005).

As outlined above, quantitative, real-time PCR can provide useful information about the viral kinetics and help guide treatment decisions. Multiple studies have shown that a VL > 10^6 copies/ml and/or a 10-fold rise in viral load are associated with progression of adenovirus infection (Claas et al., 2005; Lankester et al., 2002; Lion et al., 2003; Neofytos et al., 2007; Schilham et al., 2002; Seidemann et al., 2004). However, it is also becoming increasingly evident that there is no 'safe limit', as some reports have described patients with considerably lower viral loads in whom the infection was fatal (Lion et al., 2003). Some groups have moved to pre-emptive antiviral treatment in any patient with adenovirus detected in the blood by PCR

(Lion et al., 2003; Yusuf et al., 2006). One group has even recommended active treatment for any patient in whom adenovirus is detected, irrespective of the site of detection and the clinical features (Yusuf et al., 2006). Notably, using this approach paired with active surveillance for adenovirus, this group achieved an adenovirus-related mortality of only 1.7% in their cohort of paediatric HSCT patients.

It is also important to emphasise that there is increasing evidence that the delay of treatment is associated with poor response to antiviral therapy and consequently outcome. A report by Bordigoni et al. in HSCT patients showed that a significant delay between the detection of adenovirus and treatment initiation was statistically significantly associated with therapeutic failure (Bordigoni et al., 2001). These observations were paralleled in the study by Leruez-Ville et al., who reported that a considerable delay between the onset of symptoms and initiation of therapy (median interval 18 days) had occurred in patients with a fatal outcome, while patients who survived had overall received treatment earlier (median interval 5 days) (Leruez-Ville et al., 2004).

Based on the evidence presented here and our own experience we propose the treatment algorithm outlined in Fig. 2. However, we emphasise that the absence of data from prospective, randomised studies means definitive guidelines cannot be devised. In addition, the potential for immune recovery, as well as the other factors discussed above have to be taken into account when using the proposed guideline. The approach outlined relies on a minimum of weekly quantitative PCR on blood, in conjunction with active surveillance for adenovirus in urine and stool, as well as nasopharyngeal samples if clinically indicated. Particularly in extremely high-risk patients (e.g. paediatric HSCT, severe lymphopenia) the threshold to initiate

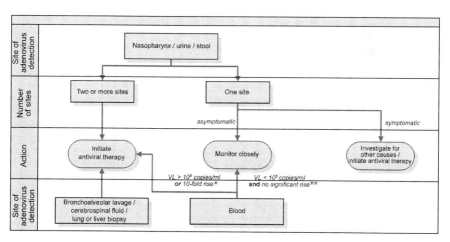

Fig. 2 Treatment algorithm for adenovirus infection in the immunocompromised host. Decisions about treatment also need to take into account the degree of immunosuppression and the likelihood of immune recovery. Particularly in paediatric HSCT patients, the threshold for treatment should be low. *This cut-off level has been shown to be associated with high mortality; therefore, at this level treatment should be started in all immunocompromised patients. **Some experts recommend treating all immunodeficient patients with adenovirus viraemia, irrespective of the viral load, or use a considerably lower threshold. A 'safe' limit has not been established.

treatment should be low, and therapy should be considered even if the arbitrary cut-off (i.e. VL < 10^5 copies/ml and less than 10-fold rise) has not been reached. Finally, in addition to antiviral therapy, other interventions should be considered early – in particular the reduction of immunosuppression if the situation allows.

10 Treatment of Adenovirus Infection

The available treatment options for adenovirus infection can be divided into immunological manoeuvres and antiviral therapy.

10.1 Immunological Interventions

Immunological manoeuvres, which include intravenous immunoglobulin (Dagan et al., 1984; Emovon et al., 2003; Seidel et al., 2003) and adoptive transfer (donor lymphocytes infusions or adenovirus-specific T cells) (Bordigoni et al., 2001; Feuchtinger et al., 2006; Hromas et al., 1994b; Leen et al., 2005; 2006) will not be discussed here, but have been reviewed elsewhere in detail (Hoffman, 2006).

10.2 Antiviral Drugs

Currently there are no antiviral drugs that are licensed for the treatment of aden-oviral infection. Also, to date, no randomised-controlled trials have been conducted in patients with adenovirus infection. Therefore, there is currently no high-quality evidence available, on which to base solid therapeutic recommendations. However, the existing literature regarding the use of cidofovir in adenovirus infection pro-vides compelling evidence that this drug is currently the most effective option in this context (Tebruegge, Clifford and Curtis; submitted).

10.2.1 Ganciclovir

Most anti-herpetic compounds, including acyclovir, penciclovir and foscarnet, have no activity against adenovirus in vitro. Ganciclovir has been shown to be the excep-tion (Naesens et al., 2005). Notably, a large study in HSCT patients showed that patients who received prophylactic or pre-emptive therapy for cytomegalovirus with ganciclovir experienced significantly lower rates of adenovirus infection (Bruno et al., 2003). While this observation indicates that ganciclovir is potentially use-ful for the prophylaxis or even treatment of adenovirus infection further data are required.

10.2.2 Vidarabine

There is conflicting data regarding the clinical effectiveness of vidarabine (*9-beta-D-arabinofuranosyladenine*) in adenovirus infection, despite the fact that activity against adenovirus has been reported in vitro (Kurosaki et al., 2004). Early reports describing the use of vidarabine for haemorrhagic cystitis in transplant patients attributed symptomatic improvement and viral clearance to treatment with this agent (Kawakami et al., 1997; Kitabayashi et al., 1994). However, other reports have described patients who continued to deteriorate clinically despite vidarabine therapy (Hatakeyama et al., 2003). Importantly, in the study in HSCT patients undertaken by Bordigioni et al., none of the seven patients who received vidarabine alone or in combination with another drug survived (Bordigioni et al., 2001).

10.2.3 Zalcitabine

Previously primarily used in the treatment of human immunodeficiency virus infection, zalcitabine (ddC/*2',3'-dideoxycytidine*) has been shown to have activity against adenovirus, both in vitro and in an animal model of adenovirus pneumonia (Mentel et al., 1997; Mentel and Wegner, 2000; Uchio et al., 2007). However, it has been pointed out that it is unlikely that adequate serum concentrations can be achieved without inducing significant mitochondrial toxicity (Naesens et al., 2005). There are currently no published clinical reports describing patients with adenovirus infection treated with this agent.

10.2.4 Ribavirin

During the early and mid 1990s there were several optimistic publications reporting cases with adenovirus infection – the majority being in transplant patients – in which ribavirin (*1-beta-D-ribofuranosyl-1,2,4-triazole-3-carboxamide*) was believed to have successfully contributed to suppression and elimination of the virus (Cassano, 1991; Liles et al., 1993; Maslo et al., 1997; Murphy et al., 1993; Sabroe et al., 1995; Wulffraat et al., 1995). This was followed by several reports describing cases in whom ribavirin failed to halt adenoviral disease, generally with fatal outcome (Chakrabarti et al., 1999; Hromas et al., 1994a; Mann et al., 1998). Further doubt was raised by the study by Bordigioni et al., in which 70% of patients receiving ribavirin as the only antiviral agent directed against adenovirus died (Bordigioni et al., 2001). Another study reported improvement in only two of 12 adenovirus-infected HSCT patients treated with intravenous ribavirin (La Rosa et al., 2001).

In a ground-breaking report published in 2005, Morfin et al showed that while ribavirin may have some activity against serotypes from species C in vitro, all other adenovirus species are intrinsically resistant to this agent (Morfin et al., 2005). This lack of in vitro activity against several adenovirus serotypes was subsequently also documented by other groups (Naesens et al., 2005).

Based on these data it could be presumed that ribavirin would be useful for the treatment of disease caused by adenovirus serotypes from species C (i.e. serotypes 1, 2, 5 & 6). However, a recent publication casts doubt on this assumption (Lankester et al., 2004). Of three patients with species C adenovirus disease (serotypes: 1, 2, 5) described in this report who received treatment with ribavirin, none were found to show a decrease in viral load using quantitative PCR.

10.2.5 Cidofovir

In vitro data shows that cidofovir has good activity against all species of adenovirus, as well as other viruses that are frequently encountered in the transplant setting, including herpes simplex virus, cytomegalovirus and Epstein-Barr virus (De Clercq, 2003; De Clercq et al., 2005; Naesens et al., 2005; Safrin et al., 1997). Early animal model studies showed the effectiveness of topical cidofovir in adenovirus kerato-conjunctivitis in rabbits (de Oliveira et al., 1996; Romanowski et al., 2001). This mode of use was more recently also explored in human subjects with some success (Hillenkamp et al., 2001; 2002).

Particularly in the HSCT setting, there is increasing evidence indicating the effectiveness of cidofovir in adenovirus disease. Notably, six studies have reported the absence of adenovirus-related fatalities in patients treated with cidofovir (Anderson et al., 2008; Bateman et al., 2006; Hoffman et al., 2001; Leruez-Ville et al., 2006; Muller et al., 2005; Nagafuji et al., 2004). This stands in stark contrast to the mortality rates between 26 and 54% reported by historical studies conducted in a similar setting (Flomenberg et al., 1994; Howard et al., 1999; La Rosa et al., 2001; Shields et al., 1985). However, the number of patients in some of these studies was relatively small, potentially skewing the data.

The two largest studies in this context were conducted by Yusuf et al. and Ljungman et al. and included 57 and 45 cases respectively (Ljungman et al., 2003;Yusuf et al., 2006). In both studies the adenovirus-related mortality in patients treated with cidofovir was low (2 and 16%, respectively) and clearance of adenovirus was achieved in the majority of patients with cidofovir therapy (98 and 65%, respectively). Although comparison with historical data is inevitably limited by the inability to control for confounding by other changes in practice, these studies provide fairly convincing evidence for the benefit of cidofovir in this setting.

There is considerably less data available on the use of cidofovir in adenovirus disease in solid organ transplant recipients (Leruez-Ville et al., 2004; Seidemann et al., 2004). Outside the transplant setting, there is even less information, with reported experience consisting only of case reports and small case series (Fianchi et al., 2003; Hedderwick et al., 1998; Hough et al., 2005; Rocholl et al., 2004; Steiner et al., 2008). However, there have been several encouraging reports (Carter et al., 2002; Keswani and Moudgil, 2007; Steiner et al., 2008; Wallot et al., 2006). Notably, one publication, which included four paediatric lung transplant recipients with adenovirus pneumonitis, reported that infection was fatal only in one patient (25%) (Doan et al., 2007). Again, this compares favourably with historical reported mortality rates in this situation of approximately 75%.

11 Future Prospects

There is compelling evidence to suggest that antiviral treatment only suppresses adenoviral replication, while immunological mechanisms in general, and the clonal expansion of adenovirus-specific T cells in particular, are required to truly control and eliminate the virus. Immunotherapy has been shown to be a promising option and its role will undoubtedly expand rapidly in the future.

The available data suggests that cidofovir is currently the most effective antiviral drug for the treatment of adenovirus infection. While more solid data from prospective, randomised, placebo-controlled trials is desirable to support this assumption, such trials would be unethical given the high levels of morbidity and mortality associated with untreated adenovirus disease in the immunocompromised host. However, it is crucial that the search for novel compounds with activity against adenovirus continues to be actively pursued, particularly as the emergence of cidofovir-resistance has recently been described in one strain of adenovirus (Kinchington et al., 2002).

References

Adeyemia, O.A., Yeldandi, A.V., & Ison, M.G. (2008). Fatal adenovirus pneumonia in a person with AIDS and Burkitt lymphoma: a case report and review of the literature. *AIDS Read,* (18), 196–198, 201–202, 206–207.

Adeyi, O.A., Randhawa, P.A., Nalesnik, M.A., Ochoa, E.R., Abu-Elmagd, K.M., Demetris, A.J., & Wu, T. (2008). Posttransplant adenoviral enteropathy in patients with small bowel transplantation. *Arch Pathol Lab Med,* (132), 703–705.

Adrian, T., Schafer, G., Cooney, M.K., Fox, J.P., & Wigand, R. (1988). Persistent enteral infections with adenovirus types 1 and 2 in infants: no evidence of reinfection. *Epidemiol Infect,* (101), 503–509.

Anderson, E.J., Guzman-Cottrill, J.A., Kletzel, M., Thormann, K., Sullivan, C., Zheng, X., & Katz, B.Z. (2008). High-risk adenovirus-infected pediatric allogenic hematopoietic progenitor transplant recipients and preemptive cidofovir therapy. *Pediatr Transplant,* (12), 219–227.

Ardehali, H., Volmar, K., Roberts, C., Forman, M., & Becker, L.C. (2001). Fatal disseminated adenoviral infection in a renal transplant patient. *Transplantation,* (71), 998–999.

Baldwin, A., Kingman, H., Darville, M., Foot, A.B., Grier, D., Cornish, J.M., Goulden, N. et al. (2001). Outcome and clinical course of 100 patients with adenovirus infection following bone marrow transplantation. *Bone Marrow Transplant,* (26), 1333–1338.

Bateman, C.M., Kesson, A.M., & Shaw, P.J. (2006). Pancreatitis and adenoviral infection in children after blood and marrow transplantation. *Bone Marrow Transplant,* (38), 807–811.

Bergelson, J.M., Cunningham, J.A., Droguett, G., Kurt-Jones, E.A., Krithivas, A., Hong, J.S., Horwitz, M.S. et al. (1997). Isolation of a common receptor for Coxsackie B viruses and adenoviruses 2 and 5. *Science,* (275), 1320–1323.

Berho, M., Torroella, M., Viciana, A., Weppler, D., Thompson, J., Nery, J., Tzakis, A., & Ruiz, P. (1998). Adenovirus enterocolitis in human small bowel transplants. *Pediatr Transplant,* (2), 277–282.

Bordigoni, P., Carret, A.S., Venard, V., Witz, F., & Le Faou, A. (2001). Treatment of adenovirus infections in patients undergoing allogeneic hematopoietic stem cell transplantation. *Clin Infect Dis,* (32), 1290–1297.

Bruno, B., Gooley, T., Hackman, R.C., Davis, C., Corey, L., & Boeckh, M. (2003). Adenovirus infection in hematopoietic stem cell transplantation: effect of ganciclovir and impact on survival. *Biol Blood Marrow Transplant,* (9), 341–352.

Cames, B., Rahier, J., Burtomboy, G., de Ville de Goyet, J., Reding, R., Lamy, M., Otte, J.B., & Sokal, E.M. (1992). Acute adenovirus hepatitis in liver transplant recipients. *J Pediatr,* (120), 33–37.

Carter, B.A., Karpen, S.J., Quiros-Tejeira, R.E., Chang, I.F., Clark, B.S., Demmler, G.J., Heslop, H.E. et al. (2002). Intravenous Cidofovir therapy for disseminated adenovirus in a pediatric liver transplant recipient. *Transplantation,* (74), 1050–1052.

Cassano, W.F. (1991). Intravenous ribavirin therapy for adenovirus cystitis after allogeneic bone marrow transplantation. *Bone Marrow Transplant,* (7), 247–248.

CDC. (2007). Acute respiratory disease associated with adenovirus serotype 14 – four states, 2006–2007. *MMWR Morb Mortal Wkly Rep,* (56), 1181–1184.

Chakrabarti, S., Collingham, K.E., Fegan, C.D., & Milligan, D.W. (1999). Fulminant adenovirus hepatitis following unrelated bone marrow transplantation: failure of intravenous ribavirin therapy. *Bone Marrow Transplant,* (23), 1209–1211.

Chakrabarti, S., Mautner, V., Osman, H., Collingham, K.E., Fegan, C.D., Klapper, P.E., Moss, P.A. et al. (2002). Adenovirus infections following allogeneic stem cell transplantation: incidence and outcome in relation to graft manipulation, immunosuppression, and immune recovery. *Blood,* (100), 1619–1627.

Claas, E.C., Schilham, M.W., de Brouwer, C.S., Hubacek, P., Echavarria, M., Lankester, A.C., van Tol, M.J. et al. (2005). Internally controlled real-time PCR monitoring of adenovirus DNA load in serum or plasma of transplant recipients. *J Clin Microbiol,* (43), 1738–1744.

Dagan, R., Schwartz, R.H., Insel, R.A., & Menegus, M.A. (1984). Severe diffuse adenovirus 7a pneumonia in a child with combined immunodeficiency: possible therapeutic effect of human immune serum globulin containing specific neutralizing antibody. *Pediatr Infect Dis,* (3), 246–251.

De Clercq, E. (2003). Clinical potential of the acyclic nucleoside phosphonates cidofovir, adefovir, and tenofovir in treatment of DNA virus and retrovirus infections. *Clin Microbiol Rev,* (16), 569–596.

De Clercq, E., Andrei, G., Balzarini, J., Leyssen, P., Naesens, L., Neyts, J., Pannecouque, C. et al. (2005). Antiviral potential of a new generation of acyclic nucleoside phosphonates, the 6-[2-(phosphonomethoxy)alkoxy]-2,4-diaminopyrimidines. *Nucleosides Nucleotides Nucleic Acids,* (24), 331–341.

de Mezerville, M.H., Tellier, R., Richardson, S., Hebert, D., Doyle, J., & Allen, U. (2006). Adenoviral infections in pediatric transplant recipients: a hospital-based study. *Pediatr Infect Dis J,* (25), 815–818.

de Oliveira, C.B., Stevenson, D., LaBree, L., McDonnell, P.J., & Trousdale, M.D. (1996). Evaluation of Cidofovir (HPMPC, GS-504) against adenovirus type 5 infection in vitro and in a New Zealand rabbit ocular model. *Antiviral Res,* (31), 165–172.

Doan, M.L., Mallory, G.B., Kaplan, S.L., Dishop, M.K., Schecter, M.G., McKenzie, E.D., Heinle, J.S. et al. (2007). Treatment of adenovirus pneumonia with cidofovir in pediatric lung transplant recipients. *J Heart Lung Transplant,* (26), 883–889.

Echavarria, M., Forman, M., Ticehurst, J., Dumler, J.S., & Charache, P. (1998). PCR method for detection of adenovirus in urine of healthy and human immunodeficiency virus-infected individuals. *J Clin Microbiol,* (36), 3323–3326.

Echavarria, M., Forman, M., van Tol, M.J., Vossen, J.M., Charache, P., & Kroes, A.C. (2001). Prediction of severe disseminated adenovirus infection by serum PCR. *Lancet,* (358), 384–385.

Edwards, K.M., Thompson, J., Paolini, J., & Wright, P.F. (1985). Adenovirus infections in young children. *Pediatrics,* (76), 420–424.

El-Sayed Zaki, M. & Abd-El Fatah, G.A. (2008). Rapid detection of oculopathogenic adenovirus in conjunctivitis. *Curr Microbiol,* (56), 105–109.

Emovon, O.E., Lin, A., Howell, D.N., Afzal, F., Baillie, M., Rogers, J., Baliga, P. et al. (2003). Refractory adenovirus infection after simultaneous kidney-pancreas transplantation: successful treatment with intravenous ribavirin and pooled human intravenous immunoglobulin. *Nephrol Dial Transplant,* (18), 2436–2438.

Erard, V., Huang, M.L., Ferrenberg, J., Nguy, L., Stevens-Ayers, T.L., Hackman, R.C., Corey, L. et al. (2007). Quantitative real-time polymerase chain reaction for detection of adenovirus after

T cell-replete hematopoietic cell transplantation: viral load as a marker for invasive disease. *Clin Infect Dis,* (45), 958–965.

Espy, M.J., Hierholzer, J.C., & Smith, T.F. (1987). The effect of centrifugation on the rapid detection of adenovirus in shell vials. *Am J Clin Pathol,* (88), 358–360.

Feuchtinger, T., Lang, P., & Handgretinger, R. (2007). Adenovirus infection after allogeneic stem cell transplantation. *Leuk Lymphoma,* (48), 244–255.

Feuchtinger, T., Matthes-Martin, S., Richard, C., Lion, T., Fuhrer, M., Hamprecht, K., Handgretinger, R., Peters, C. et al. (2006). Safe adoptive transfer of virus-specific T-cell immunity for the treatment of systemic adenovirus infection after allogeneic stem cell transplantation. *Br J Haematol,* (134), 64–76.

Fianchi, L., Scardocci, A., Cattani, P., Tartaglione, T., & Pagano, L. (2003). Adenovirus meningoencephalitis in a patient with large B-cell lymphoma. *Ann Hematol,* (82), 313–315.

Flomenberg, P., Babbitt, J., Drobyski, W.R., Ash, R.C., Carrigan, D.R., Sedmak, G.V., McAuliffe, T. et al. (1994). Increasing incidence of adenovirus disease in bone marrow transplant recipients. *J Infect Dis,* (169), 775–781.

Fox, J.P., Brandt, C.D., Wassermann, F.E., Hall, C.E., Spigland, I., Kogon, A., & Elveback, L.R. (1969). The virus watch program: a continuing surveillance of viral infections in metropolitan New York families. VI. Observations of adenovirus infections: virus excretion patterns, antibody response, efficiency of surveillance, patterns of infections, and relation to illness. *Am J Epidemiol,* (89), 25–50.

Freymuth, F., Vabret, A., Cuvillon-Nimal, D., Simon, S., Dina, J., Legrand, L., Gouarin, S. et al. (2006). Comparison of multiplex PCR assays and conventional techniques for the diagnostic of respiratory virus infections in children admitted to hospital with an acute respiratory illness. *J Med Virol,* (78), 1498–1504.

Gardner, P.S. (1968). Virus infections and respiratory disease of childhood. *Arch Dis Child,* (43), 629–645.

Garnett, C.T., Erdman, D., Xu, W., & Gooding, L.R. (2002). Prevalence and quantitation of species C adenovirus DNA in human mucosal lymphocytes. *J Virol,* (76), 10608–10616.

Greber, U.F., Suomalainen, M., Stidwill, R.P., Boucke, K., Ebersold, M.W., & Helenius, A. (1997). The role of the nuclear pore complex in adenovirus DNA entry. *Embo J,* (16), 5998–6007.

Greber, U.F., Willetts, M., Webster, P., & Helenius, A. (1993). Stepwise dismantling of adenovirus 2 during entry into cells. *Cell,* (75), 477–486.

Hale, G.A., Heslop, H.E., Krance, R.A., Brenner, M.A., Jayawardene, D., Srivastava, D.K., & Patrick, C.C. (1999). Adenovirus infection after pediatric bone marrow transplantation. *Bone Marrow Transplant,* (23), 277–282.

Hatakeyama, N., Suzuki, N., Kudoh, T., Hori, T., Mizue, N., & Tsutsumi, H. (2003). Successful cidofovir treatment of adenovirus-associated hemorrhagic cystitis and renal dysfunction after allogenic bone marrow transplant. *Pediatr Infect Dis J,* (22), 928–929.

Hedderwick, S.A., Greenson, J.K., McGaughy, V.R., & Clark, N.M. (1998). Adenovirus cholecystitis in a patient with AIDS. *Clin Infect Dis,* (26), 997–999.

Heemskerk, B., Lankester, A.C., van Vreeswijk, T., Beersma, M.F., Claas, E.C., Veltrop-Duits, L.A., Kroes, A.C. et al. (2005). Immune reconstitution and clearance of human adenovirus viremia in pediatric stem-cell recipients. *J Infect Dis,* (191), 520–530.

Hierholzer, J.C. (1992). Adenoviruses in the immunocompromised host. *Clin Microbiol Rev,* (5), 262–274.

Hillenkamp, J., Reinhard, T., Ross, R.S., Bohringer, D., Cartsburg, O., Roggendorf, M., De Clercq, E. et al. (2001). Topical treatment of acute adenoviral keratoconjunctivitis with 0.2% cidofovir and 1% cyclosporine: a controlled clinical pilot study. *Arch Ophthalmol,* (119), 1487–1491.

Hillenkamp, J., Reinhard, T., Ross, R.S., Bohringer, D., Cartsburg, O., Roggendorf, M., De Clercq, E. et al. (2002). The effects of cidofovir 1% with and without cyclosporin a 1% as a topical treatment of acute adenoviral keratoconjunctivitis: a controlled clinical pilot study. *Ophthalmology,* (109), 845–850.

Hoffman, J.A. (2006). Adenoviral disease in pediatric solid organ transplant recipients. *Pediatr Transplant,* (10), 17–25.

Hoffman, J.A., Shah, A.J., Ross, L.A., & Kapoor, N. (2001). Adenoviral infections and a prospective trial of cidofovir in pediatric hematopoietic stem cell transplantation. *Biol Blood Marrow Transplant,* (7), 388–394.

Hofland, C.A., Eron, L.J., & Washecka, R.M. (2004). Hemorrhagic adenovirus cystitis after renal transplantation. *Transplant Proc,* (36), 3025–3027.

Hong, J.Y., Lee, H.J., Piedra, P.A., Choi, E.H., Park, K.H., Koh, Y.Y., & Kim, W.S. (2001). Lower respiratory tract infections due to adenovirus in hospitalized Korean children: epidemiology, clinical features, and prognosis. *Clin Infect Dis,* (32), 1423–1429.

Hough, R., Chetwood, A., Sinfield, R., Welch, J., & Vora, A. (2005). Fatal adenovirus hepatitis during standard chemotherapy for childhood acute lymphoblastic leukemia. *J Pediatr Hematol Oncol,* (27), 67–72.

Howard, D.S., Phillips, I.G., Reece, D.E., Munn, R.K., Henslee-Downey, J., Pittard, M., Barker, M. et al. (1999). Adenovirus infections in hematopoietic stem cell transplant recipients. *Clin Infect Dis,* (29), 1494–1501.

Hromas, R., Clark, C., Blanke, C., Tricot, G., Cornetta, K., Hedderman, A., & Broun, E.R. (1994a). Failure of ribavirin to clear adenovirus infections in T cell-depleted allogeneic bone marrow transplantation. *Bone Marrow Transplant,* (14), 663–664.

Hromas, R., Cornetta, K., Srour, E., Blanke, C., & Broun, E.R. (1994b). Donor leukocyte infusion as therapy of life-threatening adenoviral infections after T-cell-depleted bone marrow transplantation. *Blood,* (84), 1689–1690.

Humar, A., Kumar, D., Mazzulli, T., Razonable, R.R., Moussa, G., Paya, C.V., Covington, E. et al. (2005). A surveillance study of adenovirus infection in adult solid organ transplant recipients. *Am J Transplant,* (5), 2555–2559.

Ison, M.G. (2006). Adenovirus infections in transplant recipients. *Clin Infect Dis,* (43), 331–339.

Jones, M.S., 2nd, Harrach, B., Ganac, R.D., Gozum, M.M., Dela Cruz, W.P., Riedel, B., Pan, C. et al (2007). New adenovirus species found in a patient presenting with gastroenteritis. *J Virol,* (81), 5978–5984.

Kalpoe, J.S., van der Heiden, P.L., Barge, R.M., Houtzager, S., Lankester, A.C., van Tol, M.J., & Kroes, A.C. (2007). Assessment of disseminated adenovirus infections using quantitative plasma PCR in adult allogeneic stem cell transplant recipients receiving reduced intensity or myeloablative conditioning. *Eur J Haematol,* (78), 314–321.

Kampmann, B., Cubitt, D., Walls, T., Naik, P., Depala, M., Samarasinghe, S., Robson, D., Hassan, A. et al. (2005). Improved outcome for children with disseminated adenoviral infection following allogeneic stem cell transplantation. *Br J Haematol,* (130), 595–603.

Kawakami, M., Ueda, S., Maeda, T., Karasuno, T., Teshima, H., Hiraoka, A., Nakamura, H. et al. (1997). Vidarabine therapy for virus-associated cystitis after allogeneic bone marrow transplantation. *Bone Marrow Transplant,* (20), 485–490.

Keswani, M. & Moudgil, A. (2007). Adenovirus-associated hemorrhagic cystitis in a pediatric renal transplant recipient. *Pediatr Transplant,* (11), 568–571.

Kim, M.R., Lee, H.R., & Lee, G.M. (2000). Epidemiology of acute viral respiratory tract infections in Korean children. *J Infect,* (41), 152–158.

Kinchington, P.R., Araullo-Cruz, T., Vergnes, J.P., Yates, K., & Gordon, Y.J. (2002). Sequence changes in the human adenovirus type 5 DNA polymerase associated with resistance to the broad spectrum antiviral cidofovir. *Antiviral Res,* (56), 73–84.

Kitabayashi, A., Hirokawa, M., Kuroki, J., Nishinari, T., Niitsu, H., & Miura, A.B. (1994). Successful vidarabine therapy for adenovirus type 11-associated acute hemorrhagic cystitis after allogeneic bone marrow transplantation. *Bone Marrow Transplant,* (14), 853–854.

Kojaoghlanian, T., Flomenberg, P., & Horwitz, M.S. (2003). The impact of adenovirus infection on the immunocompromised host. *Rev Med Virol,* (13), 155–171.

Koneru, B., Jaffe, R., Esquivel, C.O., Kunz, R., Todo, S., Iwatsuki, S., & Starzl, T.E. (1987). Adenoviral infections in pediatric liver transplant recipients. *Jama,* (258), 489–492.

Kurosaki, K., Miwa, N., Yoshida, Y., Kurokawa, M., Kurimoto, M., Endo, S., & Shiraki, K. (2004). Therapeutic basis of vidarabine on adenovirus-induced haemorrhagic cystitis. *Antivir Chem Chemother,* (15), 281–285.

La Rosa, A.M., Champlin, R.E., Mirza, N., Gajewski, J., Giralt, S., Rolston, K.V., Raad, I. et al. (2001). Adenovirus infections in adult recipients of blood and marrow transplants. *Clin Infect Dis,* (32), 871–876.

Lankester, A.C., Heemskerk, B., Claas, E.C., Schilham, M.W., Beersma, M.F., Bredius, R.G., van Tol, M.J. et al. (2004). Effect of ribavirin on the plasma viral DNA load in patients with disseminating adenovirus infection. *Clin Infect Dis,* (38), 1521–1525.

Lankester, A.C., van Tol, M.J., Claas, E.C., Vossen, J.M., & Kroes, A.C. (2002). Quantification of adenovirus DNA in plasma for management of infection in stem cell graft recipients. *Clin Infect Dis,* (34), 864–867.

Leen, A.M., Myers, G.D., Bollard, C.M., Huls, M.H., Sili, U., Gee, A.P., Heslop, H.E. et al. (2005). T-cell immunotherapy for adenoviral infections of stem-cell transplant recipients. *Ann N Y Acad Sci,* (1062), 104–115.

Leen, A.M., Myers, G.D., Sili, U., Huls, M.H., Weiss, H., Leung, K.S., Carrum, G. et al. (2006). Monoculture-derived T lymphocytes specific for multiple viruses expand and produce clinically relevant effects in immunocompromised individuals. *Nat Med,* (12), 1160–1166.

Lenaerts, L., Verbeken, E., De Clercq, E., & Naesens, L. (2005). Mouse adenovirus type 1 infection in SCID mice: an experimental model for antiviral therapy of systemic adenovirus infections. *Antimicrob Agents Chemother,* (49), 4689–4699.

Leruez-Ville, M., Chardin-Ouachee, M., Neven, B., Picard, C., Le Guinche, I., Fischer, A., Rouzioux, C. et al. (2006). Description of an adenovirus A31 outbreak in a paediatric haematology unit. *Bone Marrow Transplant,* (38), 23–28.

Leruez-Ville, M., Minard, V., Lacaille, F., Buzyn, A., Abachin, E., Blanche, S., Freymuth, F. et al. (2004). Real-time blood plasma polymerase chain reaction for management of disseminated adenovirus infection. *Clin Infect Dis,* (38), 45–52.

Leung, A.Y., Chan, M., Cheng, V.C., Yuen, K.Y., & Kwong, Y.L. (2005). Quantification of adenovirus in the lower respiratory tract of patients without clinical adenovirus-related respiratory disease. *Clin Infect Dis,* (40), 1541–1544.

Liles, W.C., Cushing, H., Holt, S., Bryan, C., & Hackman, R.C. (1993). Severe adenoviral nephritis following bone marrow transplantation: successful treatment with intravenous ribavirin. *Bone Marrow Transplant,* (12), 409–412.

Lion, T., Baumgartinger, R., Watzinger, F., Matthes-Martin, S., Suda, M., Preuner, S., Futterknecht, B. et al. (2003). Molecular monitoring of adenovirus in peripheral blood after allogeneic bone marrow transplantation permits early diagnosis of disseminated disease. *Blood,* (102), 1114–1120.

Ljungman, P., Ribaud, P., Eyrich, M., Matthes-Martin, S., Einsele, H., Bleakley, M., Machaczka, M. et al. (2003). Cidofovir for adenovirus infections after allogeneic hematopoietic stem cell transplantation: a survey by the Infectious Diseases Working Party of the European Group for Blood and Marrow Transplantation. *Bone Marrow Transplant,* (31), 481–486.

Louie, J.K., Kajon, A.E., Holodniy, M., Guardia-LaBar, L., Lee, B., Petru, A.M., Hacker, J.K. et al. (2008). Severe pneumonia due to adenovirus serotype 14: a new respiratory threat? *Clin Infect Dis,* (46), 421–425.

Mahafzah, A.M. & Landry, M.L. (1989). Evaluation of immunofluorescent reagents, centrifugation, and conventional cultures for the diagnosis of adenovirus infection. *Diagn Microbiol Infect Dis,* (12), 407–411.

Mann, D., Moreb, J., Smith, S., & Gian, V. (1998). Failure of intravenous ribavirin in the treatment of invasive adenovirus infection following allogeneic bone marrow transplantation: a case report. *J Infect,* (36), 227–228.

Maslo, C., Girard, P.M., Urban, T., Guessant, S., & Rozenbaum, W. (1997). Ribavirin therapy for adenovirus pneumonia in an AIDS patient. *Am J Respir Crit Care Med,* (156), 1263–1264.

Matar, L.D., McAdams, H.P., Palmer, S.M., Howell, D.N., Henshaw, N.G., Davis, R.D., & Tapson, V.F. (1999). Respiratory viral infections in lung transplant recipients: radiologic findings with clinical correlation. *Radiology,* (213), 735–742.

McGrath, D., Falagas, M.E., Freeman, R., Rohrer, R., Fairchild, R., Colbach, C., & Snydman, D.R. (1998). Adenovirus infection in adult orthotopic liver transplant recipients: incidence and clinical significance. *J Infect Dis,* (177), 459–462.

Medina-Kauwe, L.K. (2003). Endocytosis of adenovirus and adenovirus capsid proteins. *Adv Drug Deliv Rev,* (55), 1485–1496.

Mentel, R., Kinder, M., Wegner, U., von Janta-Lipinski, M., & Matthes, E. (1997). Inhibitory activity of 3'-fluoro-2' deoxythymidine and related nucleoside analogues against adenoviruses in vitro. *Antiviral Res,* (34), 113–119.

Mentel, R. & Wegner, U. (2000). Evaluation of the efficacy of 2',3'-dideoxycytidine against adenovirus infection in a mouse pneumonia model. *Antiviral Res,* (47), 79–87.

Michaels, M.G., Green, M., Wald, E.R., & Starzl, T.E. (1992). Adenovirus infection in pediatric liver transplant recipients. *J Infect Dis,* (165), 170–174.

Morfin, F., Dupuis-Girod, S., Mundweiler, S., Falcon, D., Carrington, D., Sedlacek, P., Bierings, M. et al. (2005). In vitro susceptibility of adenovirus to antiviral drugs is species-dependent. *Antivir Ther,* (10), 225–229.

Muller, W.J., Levin, M.J., Shin, Y.K., Robinson, C., Quinones, R., Malcolm, J., Hild, E. et al. (2005). Clinical and in vitro evaluation of cidofovir for treatment of adenovirus infection in pediatric hematopoietic stem cell transplant recipients. *Clin Infect Dis,* (41), 1812–1816.

Murphy, G.F., Wood, D.P., Jr., McRoberts, J.W., & Henslee-Downey, P.J. (1993). Adenovirus-associated hemorrhagic cystitis treated with intravenous ribavirin. *J Urol,* (149), 565–566.

Myerowitz, R.L., Stalder, H., Oxman, M.N., Levin, M.J., Moore, M., Leith, J.D., Gantz, N.M. et al. (1975). Fatal disseminated adenovirus infection in a renal transplant recipient. *Am J Med,* (59), 591–598.

Naesens, L., Lenaerts, L., Andrei, G., Snoeck, R., Van Beers, D., Holy, A., Balzarini, J. et al. (2005). Antiadenovirus activities of several classes of nucleoside and nucleotide analogues. *Antimicrob Agents Chemother,* (49), 1010–1016.

Nagafuji, K., Aoki, K., Henzan, H., Kato, K., Miyamoto, T., Eto, T., Nagatoshi, Y. et al. (2004). Cidofovir for treating adenoviral hemorrhagic cystitis in hematopoietic stem cell transplant recipients. *Bone Marrow Transplant,* (34), 909–914.

Nebbia, G., Chawla, A., Schutten, M., Atkinson, C., Raza, M., Johnson, M., & Geretti, A. (2005). Adenovirus viraemia and dissemination unresponsive to antiviral therapy in advanced HIV-1 infection. *Aids,* (19), 1339–1340.

Nemerow, G.R., Cheresh, D.A., & Wickham, T.J. (1994). Adenovirus entry into host cells: a role for alpha(v) integrins. *Trends Cell Biol,* (4), 52–55.

Neofytos, D., Ojha, A., Mookerjee, B., Wagner, J., Filicko, J., Ferber, A., Dessain, S. et al. (2007). Treatment of adenovirus disease in stem cell transplant recipients with cidofovir. *Biol Blood Marrow Transplant,* (13), 74–81.

Pacini, D.L., Collier, A.M., & Henderson, F.W. (1987). Adenovirus infections and respiratory illnesses in children in group day care. *J Infect Dis,* (156), 920–927.

Palmer, S.M., Jr., Henshaw, N.G., Howell, D.N., Miller, S.E., Davis, R.D., & Tapson, V.F. (1998). Community respiratory viral infection in adult lung transplant recipients. *Chest,* (113), 944–950.

Parizhskaya, M., Walpusk, J., Mazariegos, G., & Jaffe, R. (2001). Enteric adenovirus infection in pediatric small bowel transplant recipients. *Pediatr Dev Pathol,* (4), 122–128.

Pinchoff, R.J., Kaufman, S.S., Magid, M.S., Erdman, D.D., Gondolesi, G.E., Mendelson, M.H., Tane, K. et al. (2003). Adenovirus infection in pediatric small bowel transplantation recipients. *Transplantation,* (76), 183–189.

Rocholl, C., Gerber, K., Daly, J., Pavia, A.T., & Byington, C.L. (2004). Adenoviral infections in children: the impact of rapid diagnosis. *Pediatrics,* (113), e51–e56.

Romanowski, E.G., Yates, K.A., & Gordon, Y.J. (2001). Antiviral prophylaxis with twice daily topical cidofovir protects against challenge in the adenovirus type 5/New Zealand rabbit ocular model. *Antiviral Res,* (52), 275–280.

Rosario, R.F., Kimbrough, R.C., Van Buren, D.H., & Laski, M.E. (2006). Fatal adenovirus serotype-5 in a deceased-donor renal transplant recipient. *Transpl Infect Dis,* (8), 54–57.

Runde, V., Ross, S., Trenschel, R., Lagemann, E., Basu, O., Renzing-Kohler, K., Schaefer, U.W. et al. (2001). Adenoviral infection after allogeneic stem cell transplantation (SCT): report on 130 patients from a single SCT unit involved in a prospective multi center surveillance study. *Bone Marrow Transplant,* (28), 51–57.

Sabroe, I., McHale, J., Tait, D.R., Lynn, W.A., Ward, K.N., & Shaunak, S. (1995). Treatment of adenoviral pneumonitis with intravenous ribavirin and immunoglobulin. *Thorax,* (50), 1219–1220.

Safrin, S., Cherrington, J., & Jaffe, H.S. (1997). Clinical uses of cidofovir. *Rev Med Virol,* (7), 145–156.

Schilham, M.W., Claas, E.C., van Zaane, W., Heemskerk, B., Vossen, J.M., Lankester, A.C., Toes, R.E. et al. (2002). High levels of adenovirus DNA in serum correlate with fatal outcome of adenovirus infection in children after allogeneic stem-cell transplantation. *Clin Infect Dis,* (35), 526–532.

Schowengerdt, K.O., Ni, J., Denfield, S.W., Gajarski, R.J., Radovancevic, B., Frazier, H.O., Demmler, G.J. et al. (1996). Diagnosis, surveillance, and epidemiologic evaluation of viral infections in pediatric cardiac transplant recipients with the use of the polymerase chain reaction. *J Heart Lung Transplant,* (15), 111–123.

Seidel, M.G., Kastner, U., Minkov, M., & Gadner, H. (2003). IVIG treatment of adenovirus infection-associated macrophage activation syndrome in a two-year-old boy: case report and review of the literature. *Pediatr Hematol Oncol,* (20), 445–451.

Seidemann, K., Heim, A., Pfister, E.D., Koditz, H., Beilken, A., Sander, A., Melter, M. et al. (2004). Monitoring of adenovirus infection in pediatric transplant recipients by quantitative PCR: report of six cases and review of the literature. *Am J Transplant,* (4), 2102–2108.

Seth, P. (1994). Adenovirus-dependent release of choline from plasma membrane vesicles at an acidic pH is mediated by the penton base protein. *J Virol,* (68), 1204–1206.

Shields, A.F., Hackman, R.C., Fife, K.H., Corey, L., & Meyers, J.D. (1985). Adenovirus infections in patients undergoing bone-marrow transplantation. *N Engl J Med,* (312), 529–533.

Shirali, G.S., Ni, J., Chinnock, R.E., Johnston, J.K., Rosenthal, G.L., Bowles, N.E., & Towbin, J.A. (2001). Association of viral genome with graft loss in children after cardiac transplantation. *N Engl J Med,* (344), 1498–1503.

Sivaprakasam, P., Carr, T.F., Coussons, M., Khalid, T., Bailey, A.S., Guiver, M., Mutton, K.J. et al. (2007). Improved outcome from invasive adenovirus infection in pediatric patients after hemopoietic stem cell transplantation using intensive clinical surveillance and early intervention. *J Pediatr Hematol Oncol,* (29), 81–85.

Steiner, I., Aebi, C., Ridolfi Luthy, A., Wagner, B., & Leibundgut, K. (2008). Fatal adenovirus hepatitis during maintenance therapy for childhood acute lymphoblastic leukemia. *Pediatr Blood Cancer,* (50), 647–649.

Symeonidis, N., Jakubowski, A., Pierre-Louis, S., Jaffe, D., Pamer, E., Sepkowitz, K., O'Reilly, R.J. et al. (2007). Invasive adenoviral infections in T-cell-depleted allogeneic hematopoietic stem cell transplantation: high mortality in the era of cidofovir. *Transpl Infect Dis,* (9), 108–113.

Uchio, E., Fuchigami, A., Kadonosono, K., Hayashi, A., Ishiko, H., Aoki, K., & Ohno, S. (2007). Anti-adenoviral effect of anti-HIV agents in vitro in serotypes inducing keratoconjunctivitis. *Graefes Arch Clin Exp Ophthalmol,* (245), 1319–1325.

van Tol, M.J., Kroes, A.C., Schinkel, J., Dinkelaar, W., Claas, E.C., Jol-van der Zijde, C.M. et al. (2005). Adenovirus infection in paediatric stem cell transplant recipients: increased risk in young children with a delayed immune recovery. *Bone Marrow Transplant,* (36), 39–50.

Wallot, M.A., Dohna-Schwake, C., Auth, M., Nadalin, S., Fiedler, M., Malago, M., Broelsch, C. et al. (2006). Disseminated adenovirus infection with respiratory failure in pediatric liver transplant recipients: impact of intravenous cidofovir and inhaled nitric oxide. *Pediatr Transplant,* (10), 121–127.

Walls, T., Hawrami, K., Ushiro-Lumb, I., Shingadia, D., Saha, V., & Shankar, A.G. (2005). Adenovirus infection after pediatric bone marrow transplantation: is treatment always necessary? *Clin Infect Dis,* (40), 1244–1249.

Walls, T., Shankar, A.G., & Shingadia, D. (2003). Adenovirus: an increasingly important pathogen in paediatric bone marrow transplant patients. *Lancet Infect Dis,* (3), 79–86.

Wickham, T.J., Mathias, P., Cheresh, D.A., & Nemerow, G.R. (1993). Integrins alpha v beta 3 and alpha v beta 5 promote adenovirus internalization but not virus attachment. *Cell,* (73), 309–319.

Wigger, H.J. & Blanc, W.A. (1966). Fatal hepatic and bronchial necrosis in adenovirus infection with thymic alymphoplasia. *N Engl J Med,* (275), 870–874.

Wong, S., Pabbaraju, K., Pang, X.L., Lee, B.E., & Fox, J.D. (2008). Detection of a broad range of human adenoviruses in respiratory tract samples using a sensitive multiplex real-time PCR assay. *J Med Virol,* (80), 856–865.

Wulffraat, N.M., Geelen, S.P., van Dijken, P.J., de Graeff-Meeder, B., Kuis, W., & Boven, K. (1995). Recovery from adenovirus pneumonia in a severe combined immunodeficiency patient treated with intravenous ribavirin. *Transplantation,* (59), 927.

Yusuf, U., Hale, G.A., Carr, J., Gu, Z., Benaim, E., Woodard, P., Kasow, K.A. et al. (2006). Cidofovir for the treatment of adenoviral infection in pediatric hematopoietic stem cell transplant patients. *Transplantation,* (81), 1398–1404.

Ziring, D., Tran, R., Edelstein, S., McDiarmid, S.V., Gajjar, N., Cortina, G., Vargas, J. et al. (2005). Infectious enteritis after intestinal transplantation: incidence, timing, and outcome. *Transplantation,* (79), 702–709.

Strategies for Global Prevention of Hepatitis B Virus Infection

Pierre Van Damme, Alessandro R. Zanetti, Daniel Shouval, and Koen Van Herck

1 Hepatitis B Virus Infection

1.1 Aetiological Agent

HBV is a double-stranded, enveloped virus of the *Hepadnaviridae* family. The *Hepadna* virus family has the smallest genome of all replication competent animal DNA viruses. The single most important member of the family is HBV. Eight genotypes of HBV have been identified; they have been termed A-H. These genotypes are associated with a particular geographic distribution: genotype A being most common in the United States and Northern Europe, B and C in Asia, and D in Mediterranean countries and the Middle East (Table 1). The association of HBV genotypes with different clinical outcome and response to interferon has been described: chronic infection with genotype B appears to have a better prognosis than genotype C. Pre-core mutant infection is also most common in genotypes B, C and D, which explains why pre-core mutant infection is more common in Asia and Southern Europe. Further research is currently being performed on the role of these genotypes in transmission, disease pattern, and response to therapy (Shaefer, 2005; Erhardt et al., 2005).

The hepatitis B virion consists of a surface and a core, which contains a DNA polymerase and the e antigen. The DNA structure is double-stranded and circular with four major genes: the S (surface), the C (core), the P (polymerase), and the X (transcriptional transactivating). The S gene consists of three regions – S, pre-S1, and pre-S2 – that encode the envelope protein (HBsAg). HBsAg is a lipoprotein of the viral envelope that circulates in the blood as spherical and tubular particles. The C gene is divided into two regions, the pre-core and the core, and codes for two different proteins, the Core antigen (HBcAg) and the e antigen (HBeAg).

P. Van Damme (✉)

Faculty of Medicine, Centre for the Evaluation of Vaccination, Vaccine & Infectious Disease Institute, University of Antwerp, Antwerp, Belgium

e-mail: pierre.vanamme@ua.ac.be

A. Finn et al. (eds.), *Hot Topics in Infection and Immunity in Children VI*, Advances in Experimental Medicine and Biology 659, DOI 10.1007/978-1-4419-0981-7_14, © Springer Science+Business Media, LLC 2010

Table 1 Worldwide distribution of HBV genotypes

Genotype	Areas where genotype has been isolated
A	Africa, Europe, India, North America
B	China, Japan, Southeast Asia
C	Australia, China, Japan, Southeast Asia,
D	Mediterranean area, Middle East
E	West Africa
F	Central America, South America
G	Central America, France, South America, USA,
H	Central America, Mexico

1.2 Clinical Spectrum

Acute hepatitis B has a long incubation period (90 days on average) during this time the individual is infectious. Individual responses to the infection vary greatly, ranging from subclinical infection over a mild "flu-like" illness without jaundice to the complete clinical picture of hepatitis.

Although the acute infection is more clinically expressed in adults, infections in infants and pre-school age children are at greatest risk of becoming chronic, thereby increasing the risk of cirrhosis and primary HCC later in life, which is probably due to the effect of age on the immune system's ability to clear and eliminate the infection. About 90% of adults recover completely, although this may require 6 months or more. A small proportion (1%) of adults develop fulminant hepatitis, an exceptionally severe form of the disease, which is almost always fatal unless liver transplantation is performed (Sherlock, 1993). About 1–10% of acutely infected adults and 30–90% of infected babies will become chronically infected and remain infectious.

The course of hepatitis B virus infection is controlled by cellular and humoral immune responses. It can be tracked through serological detection of the virus particles or the antibodies raised by the immune system to target the virus. The presence of hepatitis B surface and/or hepatitis B core antibodies (anti-HBs and anti-HBc), in the absence of HbsAg, is generally taken to indicate resolution of infection and provides evidence of previous HBV infection. Persistence of HBV infection is diagnosed by the detection of HBsAg in the blood for at least 6 months or through detection of HBV-DNA, even in the absence of detectable HBsAg in patients with occult HBV infection. HBeAg is an alternatively processed protein of the pre-core gene that is only synthesized under conditions of high viral replication. Since a few years HBV-DNA has been used as an indicator for viral replication, expressed as IU/ml or copies/ml. The natural history of chronic HBV infection can vary dramatically between individuals, from a chronic carrier state (i.e., being infectious without showing any symptoms or any abnormalities on laboratory testing) over clinically insignificant or minimal liver disease without ever developing complications to clinically apparent chronic hepatitis. Chronic HBV infection can be either "replicative" (with positive HBeAg and high viral load) or "non-replicative." In the latter case, reactivation can occur either spontaneously or by immune suppression. Patients with chronic HBV and replicative infection generally have a worse prognosis and a greater chance of developing cirrhosis and/or hepatocellular carcinoma (HCC) than those without HBeAg (Chen et al., 2006). There is a clear association between serum HBV-DNA levels (viral load) and prognosis: the cumulative incidence of cirrhosis or hepatocellular carcinoma being 4.5 and 1.3%, respectively, in persons with DNA levels less than 300 copies/mL (corresponding to 50 IU/ml), while in persons with DNA levels of more than or equal to 10^6 copies/mL (corresponding to $>2 \times 10^5$ IU/ml), it is 36.2 and 14.9%, respectively. These observations provide the rationale for treating patients with high levels of HBV-DNA (Lok and McMahon, 2007; Tan and Lok, 2007).

1.3 Epidemiology and Transmission

Globally, hepatitis B is one of the most common infectious diseases. Estimates indicate that at least 2 billion people have been infected with HBV, with over 378 million people being chronic carriers (6% of the world population). Some 4.5 million new HBV infections occur worldwide each year, and 15–40% of those infected will develop cirrhosis, liver failure, or hepatocellular carcinoma (Mahoney, 1999; Lavanchy, 2004, 2005). According to the most recent World Health Organization estimate, approximately one third of all cases of cirrhosis and half of all cases of hepatocellular carcinoma can be attributed to chronic HBV infection. HBV is estimated to be responsible for 500,000–700,000 deaths each year (Shepard et al., 2006).

A model was recently developed to estimate HBV-related morbidity and mortality at country, regional, and global levels. This model calculates the age-specific risk of acquiring HBV infection, acute HBV, and progression to chronic HBV infection. HBV-related deaths among chronically infected persons were determined from HBV-related cirrhosis and HCC mortality curves, which were adjusted for background mortality. For the year 2000, the model estimated that 620,000 persons died worldwide from HBV-related causes: 94% from chronic HBV and 6% from acute HBV. Without vaccination, infections acquired during the perinatal period, in early childhood (<5 years old), and 5 years of age and older accounted for 21, 48, and 31% of HBV-related deaths, respectively (Goldstein et al., 2005)

On the basis of sero-epidemiological surveys, the World Health Organization (WHO) has classified countries into three levels of endemicity according to the prevalence of chronic HBsAg carriage: high (8% or greater), intermediate (2–8%), and low (less than 2%)

(World Health Organization, 1992, 2004). Approximately 75% of the world's chronic hepatitis B carriers live in Asian countries. China ranks highest with approximately 100 million hepatitis B carriers, and India the second highest with a carrier pool of approximately 35 million (Tandon and Tandon, 1997).

Importantly, chronic carriers of HBV are not only at risk of developing the long-term progression of the infection, but they also represent a significant source of infection to others.

HBV is transmitted by either percutaneous or mucous membrane contact with infected blood or other body fluid. The virus is found in highest concentrations in blood and serous exudates (till 10^9 virions/ml). The primary routes of transmission are perinatal, early childhood exposure (often called horizontal transmission), sexual contact, and percutaneous exposure to blood or infectious body fluids (i.e., injections, needle stick, blood transfusion).

Most perinatal infections occur among infants of pregnant women with chronic HBV infection. The likelihood of an infant developing chronic HBV infection is 70–90% for those born to HBeAg-positive mothers ($\sim$ high titers of HBV-DNA) and less than 15% for those born to HBeAg-negative mothers. Most early childhood infections occur in households of persons with chronic HBV infection. The most probable mechanism involves unapparent percutaneous or permucosal contact with infectious body fluids (e.g., bites, breaks in the skin, dermatologic lesions, skin ulcers). Recent data on paired measurements of quantitative hepatitis B virus DNA in saliva and in serum of chronic hepatitis B patients have shown the potential implication for saliva as an infectious agent (van der Eijk et al., 2004). Sexual transmission has been estimated to account for 50% of new infections among adults in industrialized countries. The most common risk factors include multiple sex partners and history of a sexually transmitted infection. Finally, in many countries, unsafe injections and other unsafe percutaneous procedures are a major source of blood-borne pathogen transmission (HBV, HCV, HIV): the risk of HBV infection from needle stick exposure to HBsAg-positive blood is approximately 30%. Worldwide unsafe injection practices account for approximately 8–16 million HBV infections each year.

In areas of high endemicity the lifetime risk of HBV infection is more than 60%, and most infections occur during the perinatal period (transmission from mother to child) or during early childhood. In areas of intermediate endemicity, the lifetime risk of HBV infection varies between 20 and 60%, and infections occur in all age groups through the four modes of transmission, but primarily in infants and children. In areas of low endemicity, infection occurs primarily in adult life by sexual or parenteral transmission (e.g., through drug use).

1.4 Prevention Strategies

All major health authorities agree that the most effective approach to reducing the burden of HBV is primary prevention of infection through universal vaccination and control of disease transmission. Because HBV-related complications mainly occur in adults, who quite often were infected with HBV as children, most of the benefits of prevention and, in particular, vaccination strategies initiated approximately 20 years ago have yet to be realized.

Interrupting the chain of infection requires knowledge of the mode of disease transmission and modification of behaviour through individual education with a focus on practicing good personal hygiene and safe sex. Screening of all donated blood and maintenance of strict aseptic techniques with invasive health treatments has reduced the likelihood of contracting HBV.

1.5 Vaccine

Since the 1980s, safe and effective HBV vaccines have been available and immunization with HBV vaccine remains the most effective means of preventing HBV disease and its consequences worldwide. Although the vaccine will not cure chronic hepatitis, it is 95% effective in preventing chronic infections from developing and is the first vaccine against a major human cancer.

After the development of plasma-derived vaccines (in 1982), which continue to be used mostly in the low and middle-income countries, recombinant DNA technology has allowed the expression of HBsAg in other organisms (Szmuness et al., 1981). As a result, different manufacturers have successfully developed recombinant DNA vaccines against HBV (commercialized in 1986).

Moreover, apart from monovalent vaccines against hepatitis B, a broad range of combination vaccines that include an HBV component exist, especially for vaccination during infancy and early childhood. Most of these simultaneously immunize against tetanus, diphtheria, and pertussis (with either a whole-cell or an acellular component); they may also include antigens for vaccination against polio and/or *H. influenzae* b. For each of these combination vaccines, it has been shown that the respective components remain sufficiently immunogenic and that the combination vaccine is safe.

More recently, so-called third-generation hepatitis B vaccines – based on the S-, preS1-, and preS2-antigens, or using new adjuvants – have been and are being developed. These vaccines specifically aim to enhance the immune response in immunocompromised persons and nonresponders (Shouval et al., 1994; Rendi-Wagner et al., 2006). Additional doses of hepatitis B vaccine (e.g., simultaneous administration of two doses at different injection sites) can elicit a seroprotective response in about half of the non-responders (Van Damme and Van Herck, 2007).

Immunization against hepatitis B requires the intramuscular administration of three doses of vaccine given at 0, 1, and 6 months. More rapid protection (i.e., for health care workers exposed to HBV or the susceptible sexual partner of a patient with acute hepatitis B) can be achieved through the adoption of an accelerated schedule using three doses of vaccine administered at 0, 1, and 2 months followed by a booster dose given at 12 months. Since becoming available, the extensive use of both plasma-derived and recombinant HBV vaccines has confirmed their safety and excellent tolerability (Niu, 1996). Side effects are generally mild, transient, and confined to the site of injection (erythema, swelling, induration). Systemic reactions (fatigue, slight fever, headache, nausea, abdominal pain) are uncommon.

However, in recent years the safety of hepatitis B vaccine has been questioned, particularly in some countries. In 1998, several case reports from France raised concern that hepatitis B vaccination may lead to new cases or relapse of multiple sclerosis (MS) or other demyelinating diseases, including Guillain-Barré syndrome. Subsequent epidemiological studies, reviewed by the Global Advisory Committee on Vaccine Safety (http://www.who.int/vaccine_safety/en/) concluded that no causal relation has been established and that the allegations were thus unfounded (Duclos, 2003). Vaccination is therefore not contraindicated in persons with a history of multiple sclerosis, Guillain-Barré syndrome, autoimmune disease (e.g., systemic lupus erythematosis or rheumatoid arthritis), or other chronic diseases. Hepatitis B vaccination is not contraindicated in pregnant or lactating women. The only absolute contraindications are known hypersensitivity to any component of the vaccine or a history of anaphylaxis to a previous dose.

Seroprotection against HBV infection is defined as having an anti-HBs level ≥ 10 IU/L after complete immunization (Szmuness et al., 1981; Centers for Disease Control and Prevention, 1987). Reviews on the use of HBV vaccine in neonates and infants report seroprotective levels of anti-HBs antibodies at 1 month after the last vaccine dose for all schedules in 98–100% of vaccinees (Safary and André, 1999; Venters et al., 2004). Another review that included studies conducted mainly in newborns reported anti-HBs levels < 10 IU/L ranging from 92.6 to 100% one month after the 0, 1, 6 months schedule and from 97 to 98% one month after an accelerated 0, 1, 2 month or 0, 1, 3 month schedule (Keating and Noble, 2003). Indeed, while HBV vaccines generally induce an adequate immune response in over 95% of fully vaccinated healthy persons, a huge interpersonal variability has been demonstrated in the immune response in healthy subjects. As such, fast/high, intermediate, slow/poor and even non-responders can be discriminated based on the magnitude and the kinetics of the immune response to HBV vaccination (Dienstag et al., 1984). The antibody response to hepatitis B vaccine has been shown to depend on the type, dosage and the schedule of vaccination used as well as on age, gender, genetic factors, co-morbidity, and the status of the immune system of the vaccinee (Hollinger, 1989; Hadler and Margolis, 1992). Immunodeficient patients such as HIV-patients or those undergoing hemodialysis or immunosuppressant therapy require higher doses of vaccine and more injections (at 0, 1, 2, and 6 months) to achieve an adequate and sustained immune response.

Follow-up studies have shown that the duration of anti-HBs positivity is related to the antibody peak level that is achieved after primary vaccination (Jilg et al., 1984; Jilg et al., 1988) vaccine-induced antibody persists over periods of at least 10–15 years. Follow-up of successfully vaccinated people has shown that the antibody concentrations usually decline over time, but clinically significant breakthrough infections are rare. Those who have lost antibody over time, after a successful vaccination, usually show a rapid anamnestic response when boosted with an additional dose of vaccine given several years after the primary course of vaccination or when exposed to the HBV. This means that the immunological memory for HBsAg can outlast the anti-HBs antibody detection, providing long-term protection against acute disease and the development of the HBsAg carrier state (West and Calandra,

1996; Banatvala and Van Damme, 2003). Hence, for immunocompetent children and adults the routine administration of booster doses of vaccine does not appear necessary to sustain long-term protection (European Consensus Group, 2000). Such conclusions are based on data collected during the first 10–20 years of vaccination in countries of both high and low endemicity (Kao and Chen, 2005; Zanetti et al., 2005).

1.6 Immunization Strategies

Since the availability of hepatitis B vaccines in industrialized countries, strategies for HBV control have stressed immunization of high-risk groups (e.g., homosexual men, health care workers, patients in sexually transmitted infection clinics, sex workers, drug users, people with multiple sex partners, household contacts with chronically infected persons, some categories of patients) and newborns to HBsAg positive mothers (along with a screening policy of pregnant women). As observed and reported in many countries, and though it is certainly desirable to immunize these persons, it is unlikely that such a program limited to high-risk groups will control HBV infection in the community (Hahné et al., 2004; van Houdt et al., 2007; Kretzschmar and de Wit, 2008). High-risk individuals are mostly difficult to target and reach; they are often infected before vaccination. Coverage of a 3-dose hepatitis B vaccine regimen also remains low in most risk groups due to low compliance and logistic reasons (Van Damme et al., 1997; Francois et al., 2002).

Furthermore, as many as 30% or more people with acute hepatitis B infection do not have identifiable risk factors and would therefore be missed by a high-risk group approach. Moreover, in the 1990s decisions to start hepatitis B immunization programs were often hampered by politicians who did not assign a high enough priority to preventive measures in public health.

In addition, most of the high endemicity countries in Africa and Asia had not made plans in the late 1980s and 1990s to introduce the hepatitis B vaccine into their Expanded Program on Immunization (EPI) schedules. The primary obstacle was the high cost of the HBV vaccine as compared with the other EPI vaccines. As the high-risk strategy made little impact on hepatitis B, while the global burden of hepatitis B disease became more and more obvious, decision makers and health planners worldwide started to discuss the strategy of universal hepatitis B immunization for a certain age cohort even those in low endemicity countries; moreover, hepatitis B is a global problem: the increasing migration and travel from and to highly endemic regions exposes more individuals to the virus and requires a global strategy (Banatvala et al., 2006; Pollard, 2007).

In 1991, the World Health Organization (WHO) called for all children to receive the HBV vaccine. Substantial progress has been made in implementing this WHO recommendation; by the end of 2007, 168 countries had implemented or were planning to implement a universal HBV immunization programme for newborns, infants (164), and/or adolescents (4) (Fig. 1). Of these 164 countries, 131 (80%) countries reported an HBV infant vaccination coverage over 80% after the third dose; these

**Countries having introduced HepB vaccine
and infant HepB3 coverage, 2006**

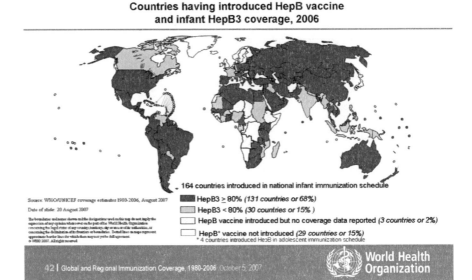

Fig. 1 Countries (*coloured*) where national infant hepatitis B vaccination has been introduced universally in the immunization schedule (2006) – infant third dose hepatitis B vaccine coverage (2006) http://www.who.int/immunization_monitoring/diseases/HepB_map_schedule.JPG

countries are mainly situated in Europe, North and South America, Northern Africa, and Australia (WHO, 2006).

High coverage with the primary vaccine series among infants has the greatest overall impact on the prevalence of chronic HBV infection in children (WHO, 2004). According to model-based predictions, universal HBV infant immunization (without administration of a birth dose of vaccine to prevent perinatal HBV infection), would prevent up to 75% of global deaths from HBV-related causes, depending on the vaccination coverage for the complete series. Adding the birth dose, would increase the proportion of deaths prevented up to 84% (Goldstein et al., 2005).

In countries with high or intermediate disease endemicity, the most effective strategy is to incorporate the vaccine into the routine infant immunization schedule or to start immunization at birth (<24 h). Countries with low endemicity may consider immunization of children or adolescents as an addition or an alternative to infant immunization (WHO, 2004, 2006; Banatvala et al., 2006)

In 2006, 44 of 53 countries in the WHO's European Region had a universal newborn, infant, childhood, and/or adolescent hepatitis B immunization program. In some very low endemic countries in Western Europe, where the HBsAg carrier rate is under 0.5%, hepatitis B is still viewed as a limited public health problem and does not warrant additional expenses on the health care budget. The United Kingdom, Ireland, the Netherlands, and the Nordic countries choose to provide hepatitis B vaccines only to well-defined risk groups in addition to screening pregnant women to identify and vaccinate exposed newborns (Banatvala et al., 2006). Changes are in view for some of these countries in the near future; for the end of 2008, the introduction of a hepatitis B infant immunization programme is planned in Ireland (Tilson et al., 2008).

1.7 Global I Impact

Countries that were early to adopt and implement universal hepatitis B immunization include Taiwan (1984), Bulgaria (1989), Malaysia (1990), the Gambia (1990), Italy, Spain, the United States (1991), and Israel (1992) (Van Damme et al., 2004; Shepard et al., 2006).

The estimated infant hepatitis B coverage increased globally from less than 1% in 1990 to 30% in 2000 to almost 50% in 2004 and to 60% in 2006 (Fig. 2) (World Health Organization. Statistics on hepatitis B. http://www.who.int/immunization_monitoring/diseases/HepB_coverage.jpg accessed on 29 February 2008). This important increase in coverage was made possible through provision of technical and financial support to 74 countries, starting in 2000, by the Global Alliance for Vaccines and Immunization (GAVI) and the Vaccine Fund, in order to address inequities in the availability of hepatitis B vaccines in developing countries. This support has clearly catalysed a tremendous increase in the number of

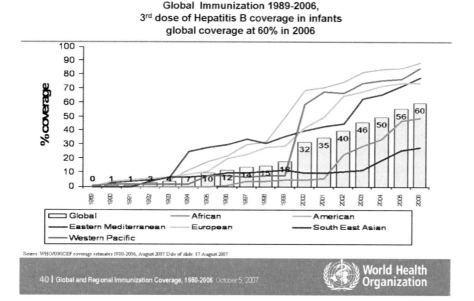

Fig. 2 Number of countries that introduced hepatitis B vaccination, and global 3° dose hepatitis B vaccine coverage for infants, 1989–2006. http://www.who.int/immunization_monitoring/diseases/hepatitis/en/index.html

economically disadvantaged countries that have introduced hepatitis B vaccine in their universal immunization programs.

The results of effective implementation of universal hepatitis B programs have become apparent in terms of reduction not only in incidence of acute hepatitis B infections but also in the carrier rate in immunized cohorts and in hepatitis-B–related mortality, which are two ways to measure the impact of a hepatitis B vaccination program (Coursaget et al., 1994; EUROHEP.NET, 2004).

Taiwan is perhaps the best example of a highly endemic area with a substantial decrease in disease burden resulting from a long-standing policy of their universal infant immunization program: the HBsAg prevalence in children younger than 15 years of age decreased from 9.8% in 1984 to 0.7% in 1999 (Chan et al., 2004). The average annual incidence of HCC among children aged 6–14 years in 1981–1986 was 0.7/100.000 while in 1990 through 1994 it was 0.36/100.000 (Chan et al., 2004). In the Gambia, since the introduction of the universal infant immunization program, childhood HBsAg prevalence decreased from 10 to 0.6% (Whittle et al., 1995; Viviani et al., 1999). In Malaysia, since implementation of a universal infant program in 1990, HBsAg seroprevalence in 7- to 12-year-old children decreased from 1.6% in 1997 to 0.3% in 2003 (Ng et al., 2005). Since the start of the infant hepatitis B vaccination program in 1991, recent data in Hawaii show a reduction of 97% in the prevalence of HbsAg. . In the period between 2002 and 2004, the incidence of new acute hepatitis B infections in children and in adults was reduced from 4.5/100.000 in 1990 to 0 (Perz et al., 2006). In Bristol Bay, Alaska, 3.2% of children were HBsAg positive before universal hepatitis B immunization; 10 years after introduction of a universal program no child under 10 years of age was HBsAg positive (Wainwright et al., 1997). Finally, in infants as well as in adolescents, surveillance data from Italy, where a universal program was started in 1991 (Piazza et al., 1988; Zanetti et al., 1993; Bonanni et al., 2003; Romanò et al., 2004; Zanetti et al., 2005), have shown a clear overall decline in the incidence of acute hepatitis B cases from 11/100,000 in 1987 to 1.6/100,000 in 2006. This decline was even more striking in people between the ages of 15–24 years in whom the morbidity rate per 100,000 fell from 17 in 1990 to less than 0.5 in 2006. Additionally, a generation of young adults (at present 27-year age cohorts) is emerging with almost no markers of HBV. In a previously hyperendemic area of South Italy (Afragola located in the greater area of Naples), the rate of HBsAg among the population was 13.4% before vaccination and dropped to 0.9% 20 years after implementation of vaccination while prevalence of anti-HBc antibody in the same population decreased from 66.9 to 7.6% (Da Villa et al., 2007). In addition, due to the biological association between HBV and hepatitis Delta virus (HDV), an added benefit is that hepatitis delta has also declined significantly in Italy after the implementation of vaccination (Gaeta et al., 2000).

Globally, in recent years, a marked progress in introduction and implementation of immunization against hepatitis B was achieved, including in countries with high prevalence of hepatitis B. In Europe, as well as in the rest of the world, there remains work to be done to support and to implement interventions that will bring us closer to the WHO goal of controlling hepatitis B in the community at large.

2 Conclusions

Despite, the availability of safe and effective vaccines and their proven effectiveness in reducing the chronic consequences of HBV infections, the current burden of disease associated with hepatitis B remains substantial. The success of vaccination programmes so far, and the interest in other vaccine-preventable diseases, have led to HBV vaccination becoming in danger of losing its place on the agenda of governments, agencies, and international organizations. To finally achieve the WHO goal of HBV elimination, continuous efforts will be required to overcome social and economic hurdles that still hamper the introduction of hepatitis B vaccination on a global scale and to keep the hepatitis B prevention on the political and public health agenda. In addition, the predicted declines in country-specific prevalences may need to take into account global migration patterns and the immigration of persons from areas that are highly HBV endemic may also need to be watched, as they may increase the disease burden within particular countries.

References

Banatvala, J.E., & Van Damme, P. (2003). Hepatitis B vaccine-do we need boosters? *J Hepatol,* (10), 1–6.

Banatvala, J.E., Van Damme, P., & Emiroglu, N. (2006). Hepatitis B immunization in Britain: time to change? *BMJ,* (332), 804–805.

Bonanni, P., Pesavento, G., Bechini, A., Tiscione, E., Mannelli, F., Benucci, C., & Nostro, A.L. (2003). Impact of universal vaccination programmes on the epidemiology of hepatitis B: 10 years of experience in Italy. *Vaccine,* (21), 685–691.

Centers for Disease Control and Prevention 1987 (1987). Recommendations of the immunization practices advisory committee. Update on hepatitis B prevention. *Morbid Mortal Weekly Report,* (36), 353–360.

Chan, C.Y., Lee, S.D., & Lo, K.J. (2004). Legend of hepatitis B vaccination: the Taiwanese experience. *J Gastroenterol Hepatol,* (19), 121–126.

Chen, C.-J., Yang, H.-I., Su, J., Jen, C.-L., You, S.-L., Lu, S.-N., Huang, G.-T. et al. (2006). Risk of HCC across a biological gradient of serum HBV-DNA levels. *JAMA,* (295), 65–73.

Coursaget, P., Leboulleux, D., Soumare, M., le Cann, P., Yvonnet, B., Chiron, J.P. et al. (1994). Twelve-year follow-up study of hepatitis B immunisation of Senegalese infants. *J Hepatol,* (21), 250–254.

Da Villa, G., Romanò, L., Sepe, A., Iorio, R., Parimbello, N., Zappa, A., & Zanetti, A.R. (2007). Impact of hepatitis B vaccination in high endemic area of south Italy and long-term duration of anti-HBs antibody in two cohort of vaccinated individuals. *Vaccine,* (25), 3133–3136.

Dienstag, J.L., Werner, B.G., Polk, B.F., Snydman, D.R., Craven, D.E., & Platt, R. (1984). Hepatitis B vaccine in health care personnel: safety, immunogenicity, and indicators of efficacy. *Ann Intern Med,* (82), 8168–8172.

Duclos, P. (2003). Safety of immunization and adverse events following vaccination against hepatitis B. *J Hepatol,* (39), 83–88.

EUROHEP.NET (2004). Data on surveillance and prevention of hepatitis A and B in 22 countries, 1990–2001. Antwerp; (also available on www.eurohep.net)

European Consensus Group on Hepatitis B immunity (2000). Are booster immunisations needed for lifelong hepatitis B immunity? *Lancet,* (355), 561–565.

Erhardt, A., Blondin, D., Hauck, K., Sagir, A., Kohnle, T., Heintges, T., & Häussinger, D. (2005). Response to interferon alfa is hepatitis B virus genotype dependent: genotype A is more sensitive to interferon than genotype D. *Gut,* (54), 1009–1013.

Francois, G., Hallauer, J., & Van Damme, P. (2002). Hepatitis B vaccination: how to reach risk groups. *Vaccine,* (21), 1–4.

Gaeta, G.B., Stroffolini, T., Chiaramonte, M., Ascione, T., Stornaiuolo, G., Lobello, S., Sagnelli, E. et al. (2000). Chronic hepatitis D: a vanishing disease? An Italian multicenter study. *Hepatology,* (32), 824–827.

Goldstein, S.T., Zhou, F., Hadler, S.C., Bell, B.P., Mast, E.E., & Margolis, H.S. (2005). A mathematical model to estimate global hepatitis B disease burden and vaccination impact. *Int J Epidemiol,* (34), 1329–1339. Available at: http://aim.path.org/en/vaccines/hepb/assessBurden/model/index.html (retrieved on 3 October 2008).

Hadler, S.C. & Margolis, H.S. (1992). Hepatitis B immunization: vaccine types, efficacy, and indications for immunization. In: Topics In Infectious Diseases. J.S. Remington, & M.N. Swartz (Eds.). (Vol. 12). Boston: Blackwell Scientific Publications, pp. 282–308.

Hahné, S., Ramsay, M., Balogun, K., Edmunds, W.J., & Mortimer, P. (2004). Incidence and routes of transmission of hepatitis B virus in England and Wales, 1995–2000: implications for immunisation policy. *J Clin Virol,* (29), 211–220.

Hollinger, F.B. (1989). Factors influencing the immune response to hepatitis B vaccine, booster dose guidelines and vaccine protocol recommendations. *Am J Med,* 87(Suppl 3A), 36–40.

Jilg, W., Schmidt, M., Zachoval, R., & Deinhardt, F. (1984). Hepatitis B vaccination: how long does protection last? *Lancet,* (2), 458.

Jilg, W., Schmidt, M., & Deinhardt, F. (1988). Persistence of specific antibodies after hepatitis B vaccination. *J Hepatol,* (6), 201–207.

Kao, J.-H. & Chen, D.S. (2005). Hepatitis B vaccination: to boost or not to boost? *Lancet,* (366), 1337–1338.

Keating, G.M. & Noble, S. (2003). Recombinant hepatitis B vaccine (Engerix-B): a review of its immunogenicity and protective efficacy against hepatitis B. *Drugs,* (63), 1021–1051.

Kretzschmar, M. & de Wit, A. (2008). Universal hepatitis B vaccination. *Lancet,* (8), 85–87.

Lavanchy, D. (2004). Hepatitis B virus epidemiology, disease burden, treatment, and current and emerging prevention and control measures. *J Viral Hepat,* (11), 97–107.

Lavanchy, D. (2005). Worldwide epidemiology of HBV infection, disease burden, and vaccine prevention. *J Clin Virol,* 34(Suppl 1), 1–3.

Lok, A.S.F. & McMahon, B.J. (2007). Chronic hepatitis B: AASLD practice guidelines. *Hepatology,* (45), 507–539.

Mahoney, F.J. (1999). Update on diagnosis, management and prevention of hepatitis B virus infection. *Clin Microbiol Rev,* (12), 351–366.

Ng, K.P., Saw, T.L., Baki, A., Rozainah, K., Pang, K.W., & Ramanathan, M. (2005). Impact of expanded programme on immunization against hepatitis B infection in school children in Malaysia. *Med Microbiol Immunol,* (194), 163–168.

Niu, M.T. (1996). Review of 12 million doses shows hepatitis B vaccine safe. *Vaccine Weekly,* (4), 13–15.

Perz, J.F., Elm, J.L., Jr., Fiore, A.E., Huggler, J.I., Kuhnert, W.L., & Effler, P.V. (2006). Near elimination of hepatitis B infections among Hawaii elementary school children universal infant hepatitis B vaccination. *Pediatrics,* (118), 1403–1408.

Piazza, M., Da Villa, G., Picciotto, L., Abrescia, N., Guadagnino, V., Memoli, A.M., Vegnente, A. et al. (1988). Mass vaccination against hepatitis B in infants in Italy. *Lancet,* (332), 1132.

Pollard, A.J. (2007). Hepatitis B vaccination. *BMJ,* (335), 950.

Rendi-Wagner, P., Shouval, D., Genton, B., Lurie, Y., Rumke, H., Boland, G., Cerny, A. et al. (2006). Comparative immunogenicity of a PreS/S hepatitis B vaccine in non- and low responders to conventional vaccine. *Vaccine,* (24), 2781–2789.

Romanò, L., Mele, A., Pariani, E., Zappa, A., & Zanetti, A. (2004). Update in the universal vaccination against hepatitis B in Italy: 12 years after its implementation. *Eur J Public Health,* 14(Suppl), S19.

Safary, A. & André, F. (1999). Over a decade of experience with the yeast recombinant hepatitis B vaccine. *Vaccine,* (18), 57–67.

Shaefer, S. (2005). Hepatitis B virus: significance of genotypes. *J Viral Hepat,* (12), 111–124.

Shepard, E.W., Simard, E.P., Finelli, L., Fiore, A.E., & Bell, B.P. (2006). Hepatitis B virus infection: epidemiology and vaccination. *Epidemiol Rev,* (28), 112–125.

Sherlock, S. (1993). Clinical features of hepatitis. In: Viral Hepatitis. A.J. Zuckerman, H.S. Thomas (Eds.). London: Churchill Livingstone, pp. 1–11.

Shouval, D., Ilan, Y., Adler, R., Deepen, R., Panet, A., Even-Chen, Z., Gorecki, M. et al. (1994). Improved immunogenicity in mice of a mammalian cell-derived recombinant hepatitis B vaccine containing pre-S_1 and pre-S_2 antigens as compared with conventional yeast-derived vaccines. *Vaccine,* (12), 1453–1459.

Szmuness, W., Stevens, C.E., Zang, E.A., Harley, E.J., & Kneller, A. (1981). A controlled clinical trial of the efficacy of the hepatitis B vaccine (Hepatavax B): a final report. *Hepatology,* (5), 377–385.

Tan, J. & Lok, A.S.F. (2007). Update on viral hepatitis: 2006. *Gastroenterology,* (23), 263–267.

Tandon, B.N. & Tandon, A. (1997). Epidemiological trends of viral hepatitis in Asia. In: Viral Hepatitis and Liver Disease. M. Rizzetto, R.H. Purcell, J.L. Gerin, & G. Verme (Eds.). Turin: Edizioni Minerva Medica, pp. 559–661.

Tilson, L., Thornton, L., O'Flanagan, D., Johnson, H., & Barry, M. (2008). Cost-effectiveness of hepatitis B vaccination strategies in Ireland: an economic evaluation. *Eur J Public Health,* 18(3), 275–282.

Van Damme, P., Kane, M., & Meheus, A. (1997). Integration of hepatitis B vaccination into national immunisation programmes. *BMJ,* (314), 1033–1037.

Van Damme, P., Van Herck, K., Leuridan, E., & Vorsters, A. (2004). Introducing universal hepatitis B vaccination in Europe: differences still remain between countries. *Eurosurveillance Weekly,* (9), 67–68.

Van Damme, P. & Van Herck, K. (2007). A review of the long-term protection after hepatitis A and B vaccination. *Travel Med Infect Dis,* (5), 79–84.

van der Eijk, A.A., Niesters, H.G.M., Götz, H.M., Janssen, H.L.A., Schalm, S.W., Osterhaus, A.D.M.E., & de Man, R.A. (2004). Paired measurements of quantitative hepatitis B virus DNA in saliva and serum of chronic hepatitis B patients: implications for saliva as infectious agent. *J Clin Virol,* (29), 92–94.

van Houdt, R., Sonder, G.J., Dukers, N.H., Bovee, L.P., van den Hoek, A., Coutinho, R.A., & Bruisten, S.M. (2007). Impact of targeted hepatitis B vaccination program in Amsterdam, the Netherlands. *Vaccine,* (25), 2698–2705.

Venters, C., Graham, W., & Cassidy, W. (2004). Recombivax-HB: perspectives past, present and future. *Expert Rev Vaccines,* (3), 119–129.

Viviani, S., Jack, A., Hall, A.J., Maine, N., Mendy, M., Montesano, R., & Whittle, H.C. (1999). Hepatitis B vaccination in infancy in the Gambia: protection against carriage at 9 years of age. *Vaccine,* (17), 2946–2950.

Wainwright, R., Bulkow, L.R., Parkinson, A.J., Zanis, C., & McMahon, B.J. (1997). Protection provided by hepatitis B vaccine in a Yupik Eskimo Population: results of a 10 year study. *J Infect Dis,* (175), 674–677.

West, D.J. & Calandra, G.B. (1996). Vaccine induced immunologic memory for hepatitis B surface antigen: implications for policy on booster vaccination. *Vaccine,* (14), 1019–1027.

Whittle, H.C., Maine, N., Pilkington, J., Mendy, M., Fortuin, M., Bunn, J., Allison, L. et al. (1995). Long-term efficacy of continuing hepatitis B vaccination in infancy in two Gambian villages. *Lancet,* (345), 1089–1092.

World Health Organization 1992. (1992). Informal consultation on quadrivalent diphtheria-tetanus-pertussis-hepatitis B vaccine. Final Report, pp.1–12, Geneva.

World Health Organization (2004). Hepatitis B vaccines (WHO position paper). *Weekly Epidemiol Record,* (79), 255–263.

World Health Organization 2006. Vaccines and biologicals. WHO vaccine preventable disease monitoring system. Global summary 2006 (data up to 2005). Available at http://www.who.int/vaccines-documents/GlobalSummary.pdf (retrieved on 15 September 2007).

Zanetti, A.R., Tanzi, E., Romanò, L., & Grappasonni, I. (1993). Vaccination against hepatitis B: the Italian strategy. *Vaccine,* (11), 521–524.

Zanetti, A.R., Mariano, A., Romanò, L., D'Amelio, R., Chironna, M., Coppola, R.C., Cuccia, M., Mangione, R., Marrone, F., Negrone, F.S., Parlato, A., Zamparo, E., Zotti, C., Stroffolini, T., & Mele, A; Study Group. (2005). Long-term immunogenicity of hepatitis B vaccination and policy for booster: an Italian multicentre study. *Lancet,* 366(9494), 1379–1384.

Subject Index

A. Finn et al. (eds.), *Hot Topics in Infection and Immunity in Children VI,* Advances
in Experimental Medicine and Biology 659, DOI 10.1007/978-1-4419-0981-7,
© Springer Science+Business Media, LLC 2010